高等学校教材

实用计算机英语阅读教程

明净 朱晓荣 主 编
赵丹 吴玲敏 副主编

化学工业出版社
·北京·

内容简介

《实用计算机英语阅读教程》主要是针对普通高等院校，尤其是二本（独立）院校工科计算机科学大类（包括计算机科学与技术、数字媒体技术、网络工程、软件工程、物联网工程、智能科学与技术、信息安全、电子与计算机工程、空间信息与数字技术等）学生开展的专业英语导入式的阅读学习。本教材内容一共包含10个单元：前4个单元为大学英语阅读策略技巧讲解，每单元在进行技巧讲解的同时配以计算机科学相关文章阅读题，实现讲练结合；后6个单元为计算机科学技术相关主题阅读，包括网络技术、人工智能（AI）技术、5G技术、虚拟现实（VR）技术、物联网、数据安全，每一个单元包含与该主题一致且有难度区分的三篇英文阅读文章，第一篇为单元主题课文，第二、三篇为拓展阅读文章，其中一篇展示我国在该技术领域取得的研究成果或研究现状，在单元课文结尾附有文化词条、重难点词汇及课后习题，每篇阅读文章都辅以一定量的阅读练习。

该教材的目标是帮助计算机科学大类相关专业的学生掌握英语阅读策略，积累专业相关的英语词汇，从而能阅读专业英语文章，为其后续更深入的专业英语学习和职场所需英文阅读能力提供保障。

图书在版编目（CIP）数据

实用计算机英语阅读教程 / 明净，朱晓荣主编；赵丹，吴玲敏副主编. -- 北京：化学工业出版社，2024.12. -- ISBN 978-7-122-47093-5

Ⅰ．TP3

中国国家版本馆CIP数据核字第2024RQ4624号

责任编辑：李 琰	装帧设计：韩 飞
责任校对：宋 夏	

出版发行：化学工业出版社（北京市东城区青年湖南街13号　邮政编码100011）
印　　装：涿州市般润文化传播有限公司
787mm×1092mm　1/16　印张11¼　字数253千字　2025年3月北京第1版第1次印刷

购书咨询：010-64518888　　　　　　　　　　　售后服务：010-64518899
网　　址：http://www.cip.com.cn
凡购买本书，如有缺损质量问题，本社销售中心负责调换。

定　　价：38.00元　　　　　　　　　　　　　　版权所有　违者必究

前言

大学英语因为坚持通用英语课程教学模式而饱受争议。以通用英语能力训练为基础的教学模式无法满足本科生对专业英语学习的需求，也无法满足职场要求。大学英语在国家实施创新人才培养战略的大背景下必须尽快从通用英语范式向专用英语范式转移。近年来大学英语增添了诸如英美概况、中国文化概论、高级英语、跨文化交际、职场英语、高级英语写作等课程，以满足已经达到课程教学基本要求的非英语专业本科学生的需求，但效果依然不理想，这是因为大学英语本质上依然延续通识外语教育模式。大学英语后续课程内涵和外延的教育改革已经是大势所趋。课程改革必须以教学过程中占据重要地位的教材改革为先行，因为教材是课程的重要载体和开展教学的主要抓手。

此外，现行的行业英语教材涉及商贸、旅游、医学、法律等领域，根据其专业特性，这几类英语教材的课文定位及内容选择基本以人物对话为主，旨在教授专业词汇的同时也培养学习者在这类职业的特定场景中的情景交际能力。而计算机科学大类相关专业与上述专业相比，具有其自身特殊性，从事计算机行业的专业技术人员的工作并不以人与人的交际沟通、询问回答为主，而着重于对各种计算机科学相关文本的阅读理解、各种计算机科学术语的认知等。

《实用计算机英语阅读教程》正是基于以上情况编写而成。作者在编写过程中，从阅读策略的理论阐述，到主题单元的选择、阅读练习配备，再到教学方法，都力求突出"实用"二字。

目前市面上的计算机英语教材大多是非常专业的行业英语，即以英语书写的专业教材，具备专业知识的广度和深度，此类教材难度较大，给学生和老师都提出了很大的挑战。对于学生而言，至少是通过国家英语4、6级考试，或是准备读研深造，有行业学术英语的阅读需求。对于教师而言，则是需要英语能力好的计算机科学专业教师。而《实用计算机英语阅读教程》则是一本能够衔接通识英语学习和计算机专业英语学习的行业英语教材，可由英语教师实施教学，重在帮助学生提高行业英语阅读能力、积累专业英语学术词汇。

本教材一共有10个单元，前4个单元为大学英语阅读策略技巧讲解，包括：略读和寻读、理解文章结构、明确作者的写作意图和态度、词义和句意推测，每单元分为两章，每章都按照讲练结合的模式，分为策略讲解、范文阅读、阅读演练三节内容，选取的阅读材料大多与计算机科学技术相关。后6个单元为计算机科学技术相关主题阅读，包括网络技术、人工智能（AI）技术、5G技术、虚拟现实（VR）技术、物联网、数据安全，每单元分为两章：第一章为主题阅读课文及练习，课文选择能够反映该计算机技术领域国外研究发展现状的阅读材料，练习包括课文阅读理解问答题、词汇填空题、翻译题和主题阅读写作题；第二章为拓展阅读，提供2篇阅读材料，一篇反映我国在该技术领域的发展现状，与课文进行对照学习，另一篇则为课文主题相关的阅读材料，两篇阅读均配有阅读理解题。这10个单元可满足40个学时的英语阅读课程教学需求。

其中，明净负责第一～第三单元、第六单元的编写及全书内容的编排审核，朱晓荣负责第四、五、九、十单元的编写，吴玲敏负责第七单元的编写，赵丹负责第八单元的编写。

由于编者的水平和经验有限，书中难免会有疏漏及不足之处，恳请使用本书的师生和其他读者批评指正。

明净

2024年1月25日于武汉

目 录

第一单元　阅读技巧：略读和寻读 `001`

　　第一章　略读寻找主题大意 ·················002
　　　　第一节　策略讲解 ·····················002
　　　　第二节　范文阅读 ·····················004
　　　　第三节　阅读演练 ·····················007

　　第二章　寻读查找具体信息 ·················011
　　　　第一节　策略讲解 ·····················011
　　　　第二节　范文阅读 ·····················014
　　　　第三节　阅读演练 ·····················018

第二单元　阅读技巧：理解文章结构 `023`

　　第一章　明确主要支撑细节 ·················025
　　　　第一节　策略讲解 ·····················025
　　　　第二节　范文阅读 ·····················026
　　　　第三节　阅读演练 ·····················032

　　第二章　区分事实和观点 ···················035
　　　　第一节　策略讲解 ·····················035
　　　　第二节　范文阅读 ·····················037
　　　　第三节　阅读演练 ·····················039

第三单元　阅读技巧：明确作者的意图和态度 `042`

　　第一章　明确作者的写作意图 ···············043
　　　　第一节　策略讲解 ·····················043
　　　　第二节　范文阅读 ·····················044
　　　　第三节　阅读演练 ·····················047

第二章　明确作者的观点态度 ·· 051
　　第一节　策略讲解 ·· 051
　　第二节　范文阅读 ·· 053
　　第三节　阅读演练 ·· 057

第四单元　阅读技巧：词义和句意推测　　　　　　　　　060

第一章　词义推测 ·· 061
　　第一节　策略讲解 ·· 061
　　第二节　范文阅读 ·· 064
　　第三节　阅读演练 ·· 066

第二章　句意推测 ·· 069
　　第一节　策略讲解 ·· 069
　　第二节　范文阅读 ·· 071
　　第三节　阅读演练 ·· 076

第五单元　网络技术　　　　　　　　　　　　　　　　　080

第一章　主题阅读 ·· 081
　　Text　The Internet's Next Act ································· 081
第二章　拓展阅读 ·· 087
　　Reading 1　Digital Economy Expedites Quality Development ······ 087
　　Reading 2　Cloudification Will Mean Upheaval in Telecoms ······ 090

第六单元　人工智能技术　　　　　　　　　　　　　　　094

第一章　主题阅读 ·· 095
　　Text　AI's New Frontier ·· 095
第二章　拓展阅读 ·· 103
　　Reading 1　AI's Role in Driving Growth Bigger ················· 103
　　Reading 2　Could OpenAI be the Next Tech Giant? ··············· 105

第七单元　5G 技术　　　　　　　　　　　　　　　　　　112

第一章　主题阅读 ·· 113
　　Text　How 5G Can Unlock the Potential of Smart Homes ·········· 113

第二章　拓展阅读 ··· 121
　　　　Reading 1　China Builds World's Largest 5G Network with 475 Million
　　　　　　　　　　Users ·· 121
　　　　Reading 2　Will the Cloud Business Eat the 5G Telecoms Industry? ··· 124

第八单元　虚拟现实技术　　　　　　　　　　　　　　　　　　　　128

　　第一章　主题阅读 ··· 129
　　　　Text　VR Continues to Make People Sick - and Women More So than
　　　　　　　Men ·· 129
　　第二章　拓展阅读 ··· 137
　　　　Reading 1　Tech Giants Bank on VR for Metaverse Opportunities ······ 137
　　　　Reading 2　Metaverse No More? ByteDance and Tencent Scale back
　　　　　　　　　　VR Ambitions ··· 139

第九单元　物联网　　　　　　　　　　　　　　　　　　　　　　　　143

　　第一章　主题阅读 ··· 144
　　　　Text　The Internet of Things: Applications for Business ··············· 144
　　第二章　拓展阅读 ··· 150
　　　　Reading 1　China Standard Plan 2035 ··· 150
　　　　Reading 2　China Releases 10-year Vision, Action Plan for BRI, Focusing
　　　　　　　　　　on Green, Digital Development and Supply Chain ··········· 153

第十单元　数据安全　　　　　　　　　　　　　　　　　　　　　　　157

　　第一章　主题阅读 ··· 158
　　　　Text　Cybercriminals Are Now Targeting Top Executives and
　　　　　　　Could Be Using Sensitive Information to Extort Them ············· 158
　　第二章　拓展阅读 ··· 164
　　　　Reading 1　How Didi Crashed Into China's New Data Security Laws··· 164
　　　　Reading 2　Data Security and GDPR ··· 167

Unit One

第一单元

阅读技巧：略读和寻读

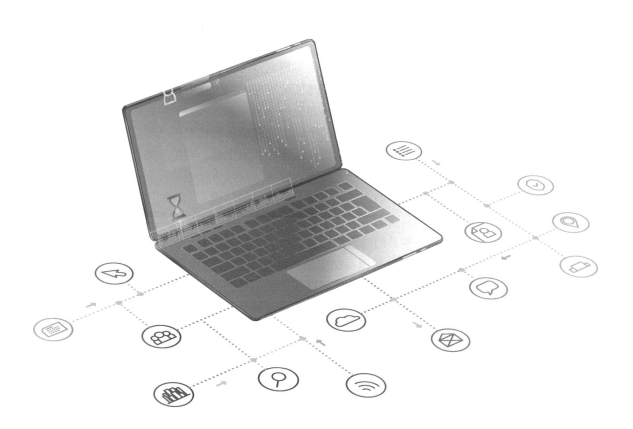

在日常生活中我们通常会运用不同的策略和方法阅读不同类型的文章。我们可以通过略读（skimming）把握文章大意，通过寻读（scanning）确定细节信息，或者通过精读详细理解信息。我们的阅读方式取决于文章的长度、类型以及阅读目的。

第一章　略读寻找主题大意

第一节　策略讲解

略读（skimming）是常用的阅读方法之一。Skimming 意为 skim a text quickly to understand the main idea，顾名思义，就是粗略地阅读或者跳跃式阅读，其主要特征是在有限的时间内选择性地阅读——筛选有用信息，省略无关信息。通常的阅读要求是阅读每一个单词，每次注目看 1~2 个词。而略读作为一种速读（大约每分钟 100~150 个词），其目的不在于精确地理解文章，而是在尽可能短的时间内总结一篇文章的主旨大意和脉络，暂时不关注其细节。我们可以利用文章标题、副标题或段落主旨句和关键词等一系列的方式辨识要点，迅速有效地获取重要信息，以提高阅读速度。当然我们也要尽可能地去理解文章内容。一般来说，略读的速度比普通阅读速度快 3 到 4 倍。

略读作为一种快速阅读技巧对我们来说并不陌生。这就像我们看报纸时，几版、十几版的页面不可能一字不漏地往下看，只能用最快的速度找出主题，略掉一些无关紧要的细节或与主题不甚相干的部分，知道其中的大概内容即可。这种方法要求我们不要把眼睛盯在具体的单词上，而是要看篇章中的主题句或者是从篇章的结构着手，利用自己的推理能力，对文章的信息进行分析，从而归纳总结出主题。善于略读的人会有选择地阅读，跳过一些无关的句子、段落甚至是整页的内容，只看自己感兴趣的内容。

略读的大体步骤是：先读文章标题及文中小标题（如果有），再读段落的开头和结尾部分，或者文章的第一段和最后一段，这样可以帮助我们很快找到文章的主旨大意。

一篇文章的主旨往往会在文章的第一段或开头局部找到，作者会通过"首段提问法""转移重心法""独句段开头法""类比开头法""事例开头法"等方法暗示或引入主题。

如果文章第一段的第一句、最后一句或者第二段第一句是一个疑问句，那么本文的主旨就很可能是对该问题的一个解答。如果文章第一段先描述一件事情，然后在本段或第二段首句出现转折词，进而表述另一件事情，那么本文主旨就是转折词之后的这件事。如果文章开头通过引用、表达个人经历、讲述故事、描写一项研究等方式举例，其目的明显是要通过举例的方法引出文章主题。通常情况下，这类文章的主题都会在首段末句或第二段首句。如果一篇文章开头或第一段涉及一项内容 A，随后或者第二段开头用到"Now the same thing is happening to B"等表示类比的句型，这时可以确定本文主旨是 B 内容。如果一篇文章的第一段是一句话，那么本文的主旨就是本句话所表达的内容。如果作者在文章的开头引用社会公认的观点或者他人的观点，其目的显然是要通过这些观点引出自己的观点。通常情况下，这类文章的主题会在表述社会公认观点或他人观点的句子之后。有些文章的第一段可能并没有采用上述方法来引入主题，这时可以考虑本篇文章是否采用了"各段分述，围绕主题"的方法。同理，如果想要抓住某段的大意，也是通过阅读该段的前两句和后两句。

采用略读可以帮助学生迅速通读全文，在这个过程中把握文章或者段落的主旨、掌握大致结构，进而提炼段落内部层次，快速找到关键信息的对应位置，从而提炼出所需的信息。

总之，略读的关键在于一个"略"字，注意力应该集中在主题，而不是细节上。试以下文为例进行略读。

Three passions, simple but overwhelmingly strong, have governed my life: the longing for love, the search for knowledge, and unbearable pity for the suffering of mankind. These passions, like great winds, have blown me here and there.

I have sought love, first because it brings ecstasy so great that I would often have sacrificed all the rest of life for a few hours of this joy. I have sought it, next, because it relieves loneliness that terrible loneliness in which one shivering consciousness looks over the rim of the world into the cold unfathomable lifeless abyss. I have sought it, finally, because in the union of love I have seen, in a mystic miniature, the vision of the heaven that saints and poets have imagined. This is what I sought, and though it might seem too good for human life, this is what at last I have found.

With equal passion I have sought knowledge. I have wished to understand the hearts of men. I have wished to know why the stars shine. And I have tried to apprehend the Pythagorean power by which number holds sway above the flux.

Love and knowledge, so far as they were possible, led upward toward the heavens. But always pity brought me back to earth. Echoes of cries of pain reverberated in my heart. Children in famine, victims tortured by oppressors, helpless old people a hated burden to their sons, and the whole world of loneliness, poverty, and pain make a mockery of what human life should be. I long to alleviate the evil, but I cannot, and I too suffer.

This has been my life, I have found it worth living, and would gladly live it again if the chance were offered me.

使用略读法，我们的目光只去捕捉以下关键的字眼：

第一段：**Three passions... have governed my life**: (l) the longing for love, (2) the search for knowledge, (3) unbearable pity for the suffering of mankind;

第二段：**I have sought love because**: (l) it brings ecstasy…, (2) it relieves loneliness…, (3) I have seen…the vision of the heaven;

第三段：**With equal passion I have sought knowledge. I have wished to** (l) understand…, (2) know…(3) **And** I have tried to apprehend …

第四段：Love and knowledge…led upward toward the heavens. **But** always pity brought me back to earth. …I long to alleviate the evil, **but** I cannot, and I too suffer.

第五段：This has been my life…and would gladly live it again.

这样一来，大大减轻了阅读的负担，一篇近 300 词的文章，阅读其中的 100 个词就能概括全部的内容。略读归纳起来也就两句话：**抓主谓结构，看前后衔接**。

第二节 范文阅读

Passage One

Skim the following passage and answer the questions in 2 minutes.

After a day's work, you are not quite ready to go home. Perhaps you fancy catching a film. You could head to the cinema. Instead, you retreat into your car. A few taps on the touchscreen dashboard and the vehicle turns into a multimedia cocoon. Light trickles down the interior surfaces like a waterfall. Speakers ooze surround-sound. Augmented-reality glasses make a screen appear in front of your eyes. This immersive experience is at the core of what Nio, a Chinese electric-vehicle (EV) company, laid out as the future of the car at a launch party in October in Berlin. "We have a supercomputer in our cars," boasts Nio's boss, William Li.

(111 words)

计算机词汇：

dashboard n. (汽车的) 仪表板	electric-vehicle (EV) n. 电动汽车
touchscreen n. 触摸屏	augmented-reality phr. 增强实境

Ex. Which of the following is the passage talk about?

　　A) Enjoying a film after work in the cinema.

　　B) Testing the touchscreen dashboard.

　　C) Experiencing an immersive film-watching in a car.

　　D) Installing a supercomputer in an electric vehicle.

【答案与解析】C. 本篇节选是关于蔚来汽车的智能化体验。全篇共 10 句，前 8 句属于描述性句子，第 9 句对前面的场景描述进行总结，是本段的主旨大意。

Passage Two

Skim the following passage and answer the questions in 3 minutes.

Many people think that information technology and biotechnology will rule the 21st century. Robert Birge, a chemist at the University of Connecticut, is trying to combine them, by making computer memories out of protein (蛋白质).

The protein in question is bacteriorhodopsin (bR) (细菌视紫红质), a molecule that undergoes a structural change when it absorbs light. By using genetic engineering to tweak (扭) wild bR from a bacterium (细菌) called Halobacterium salinarum, Dr Birge and his colleagues have made a variety of the molecule that they claim is well-suited to act as an element of a computer's memory. Hit with a green light, it adopts one shape. Hit subsequently with a red light, it twists itself into another. Then, if hit with blue light, it resets itself into its original state.

(125 words)

计算机词汇：
information technology	信息技术	biotechnology	n. 生物技术
genetic engineering	基因工程	computer memory	计算机存储器

Ex. What is the passage mainly about?

A) What will rule the 21st century.

B) How bacterial protein can be used in computers.

C) What we can get from bacterial protein.

D) How to make bacterial protein.

【答案与解析】B. 本文论述了信息技术和生物技术如何完美地结合在用蛋白质做电脑存储器这一新技术当中。第一段即引出论点，为全文的主旨大意段。

Passage Three

Skim the following passage and answer the questions in 6 minutes.

The "paperless office" has earned a proud place on lists of technological promises that did not come to pass. Surely, though, the more modest goal of the carbon-paperless office is within the reach of mankind? Carbon paper allows two copies of a document to be made at once. Nowadays, a couple of keystrokes can do the same thing with a lot less fuss.

Yet carbon paper persists. Forms still need to be filled out in a way that produces copies. This should not come as a surprise. Innovation tends to create new niches (合适的职业), rather than refill those that already exist. So technologies may become marginal, but they rarely go extinct. And today the little niches in which old technologies take refuge are ever more viable and accessible, thanks to the Internet and the fact that production no longer needs to be so mass; making small numbers of obscure items is growing easier.

On top of that, a widespread technology of nostalgia (技术怀旧) seeks to preserve all the ways people have ever done anything, simply because they are kind of neat. As a result, technologies from all the way back to the stone age persist and even fourish in the modern world. According to *What Technology Wants*, a book by Kevin Kelly, one of the founders of Wired magazine, America's flintknappers (燧石) produce over a million new arrow and spear heads every year. One of the things technology wants, it seems, is to survive.

Carbon paper, to the extent that it may have a desire for self-preservation, may also take comfort in the fact that, for all that this is a digital age, many similar products are hanging on, and even making comebacks. Indeed, digital technologies may prove to be more transient than their predecessors. They are based on the idea that the medium on which a file's constituent 0s and 1s are stored doesn't matter, and on Alan Turing's insight that any computer can mimic any other, given enough memory and time. This suggests that new digital technologies should be able to wipe out their predecessors completely. And early digital technologies do seem to be vanishing. The music cassette is enjoying a little renaissance, its very faithlessness apparently part of its charm; but digital audio tape seems doomed.

So revolutionary digital technologies may yet discard older ones to the dustbin. Perhaps this will be the case with a remarkable breakthrough in molecular (分子的) technology that could, in principle, store all the data ever recorded in a device that could fit in the back of a van. In this instance, it would not be a matter of the new extinguishing the old. Though it may never have been used for MP3s and PDFs before, DNA has been storing data for over three billion years. And it shows no sign of going extinct.

(487 words)

计算机词汇：
carbon-paperless office 碳素无纸化办公室　　　　keystroke *n.& v.* 击键；用键盘输入
constituent　*n.* 构成要素　　　　　　　　　　　Alan Turing [人名] 阿兰图灵
store n. (计算机的)存储器；v. (在计算机中)存储　digital adj. 数字的

Ex.1. Underline the topic sentence(s) of each paragraph with a line.

Ex.2. The passage mainly concerned with_____.

　　A) the difficulty of the realization of paperless office.

B) the fact that newest technologies may die out while the oldest survive.

C) the reason why old technologies will never be on the edge of extinction.

D) the importance of keeping improving technologies all the time.

【答案与解析】

Ex.1. Para 1: The "paperless office" has earned a proud place on lists of technological promises that did not come to pass.

Para 2: Yet carbon paper persists.

Para 3: On top of that, a widespread Technology of nostalgia seeks to preserve all the ways people have ever done anything, simply because they are kind of neat.

Para 4: Carbon paper, to the extent that it may have a desire for self-preservation, may also take comfort in the fact that, for all that this is a digital age, many similar products are hanging on, and even making comebacks.

Para 5: So revolutionary digital technologies may yet discard older ones to the dustbin.

【注】虽然第五段的主旨句提到了数字技术可能会淘汰旧技术，但后文通过DNA存储数据的例子暗示了并非所有旧技术都会被淘汰，而是强调了某些技术的持久性。然而，从段落的整体内容来看，该段主要讨论的是数字技术可能带来的变革和对旧技术的影响，因此上述主旨句仍然适用。

Ex.2. B. 本题考查全文主旨，应对全文进行整体把握。文章以复写纸为例说明旧技术不会消亡，接下来论证其原因，最后表明最新的技术看起来最有可能消失；而最古老的技术有可能一直与我们相伴，故答案为B）。A）"无纸办公室实现的困难性"仅在文中第一段提及，故排除；C）"为什么旧技术永远不会消亡"仅在文中第二、三段涉及，故排除；D）"不断更新技术的重要性"是对原文的曲解，故排除。

第三节 阅读演练

Passage One

Skim the following passage and answer the question 1 in five minutes.

There is no denying that students should learn something about how computers work, just as we expect them at least to understand that the internal-combustion engine (内燃机) has something to do with burning fuel, expanding gases and pistons (活塞) being driven. For people should have some basic idea of how the things that they use do what they do. Further, students might be helped by a course that considers the computer's impact on society. But that is not what is meant by computer literacy. For computer literacy is not a form of literacy (读写能力); it is a trade skill that should not be taught as a liberal art.

Learning how to use a computer and learning how to program are two distinct activities. A case might be made that the competent citizens of tomorrow should free themselves from their fear of computers. But this is quite different from saying that all ought to know how to program. Leave that to people who have chosen programming as a career. While programming can be lots of fun, and while our society needs some people who are experts at it, the same is true of auto repair and violin-making.

Learning how to use a computer is not that difficult, and it gets easier all the time as programs become more "user-friendly". Let us assume that in the future everyone is going to have to know how to use a computer to be a competent citizen. What does the phrase "learning to use a computer" mean? It sounds like "learning to drive a car", that is, it sounds as if there is some set of definite skills that, once acquired, enable one to use a computer.

In fact, "learning to use a computer" is much more like "learning to play a game", but learning the rules of one game may not help you play a second game, whose rules may not be the same. There is no such a thing as teaching someone how to use a computer. One can only teach people to use this or that program and generally that is easily accomplished.

(361 words)

计算机词汇：

computer literacy 计算机知识；有使用计算机的能力

program *n.* (计算机) 程序 (=programme) *v.* (给计算机) 编写程序，设计程序

programming *n.* 编程   programming language 编程语言

user-friendly *adj.* 用户友好的；界面友好的

Question1: The author's purpose in writing this passage is _____.

A) to stress the impact of the computer on society.

B) to explain the concept of computer literacy.

C) to illustrate the requirements for being competent citizens of tomorrow.

D) to emphasize that computer programming is an interesting and challenging job.

Passage Two

Skim the following passage and answer the question 2 in six minutes.

Para 1: Artificial intelligence is here — and it's bringing new possibilities, while also raising questions. AI and big-data technology will transform the world.

Para 2: Companies aim to make a complicated world simpler through technology. Companies can develop products and services that improve people's lives. They will be more modest machines that will drive your car, translate foreign languages, organize your photos, recommend entertainment options and maybe diagnose your illness.

Para 3: Baidu's speech-recognition software — which can accomplish the difficult task of hearing tonal difference in Chinese dialects — is considered top of the class. Think of a spreadsheet on steroids, trained on big data. These tools can outperform human beings at a given task.

Para 4: New technologies have enabled people to be more productive, as the Internet and word processing have for office workers. More productive workers, in turn, earn more money and produce goods and services than improve lives.

Para 5: The A I products that now exist are improving faster than most people realize and promise to radically transform our world, not always for the better.

Para 6: There are fears that technology will displace/eliminate jobs and lead to mass unemployment. The technological breakthroughs of recent years — allowing machines to mimic the human mind — are enabling machines to do knowledge jobs and service jobs, in addition to factory and clerical work. Machines are mastering ever more intricate tasks, such as translating texts or diagnosing illnesses.

Para 7: Unlike the Industrial Revolution and the computer revolution, the A I revolution is not taking certain jobs (artisans, personal assistants who use paper and typewriters) and replacing them with other jobs (assembly-line workers, personal assistants conversant with computers). but instead it would create as many jobs as it destroyed. Although some tasks are automatable, many others rely on human judgment.

Para 8: There are certain human skills machines will probably never replicate, like common sense, adaptability and creativity.

(306 words)

Question 2: Matching the paragraphs with the corresponding headings below.

A) The world will be transferred by the A I products.

B) The purpose of the company is to simplify the complicated world with technology.

C) A I revolution will not take some jobs since some jobs still requires human.

D) People are afraid that machines will replace human beings.

E) Humans have some unique skills which cannot be mimic by machines.

F) There arise some questions that A I technology will change the world.

G) New technologies will bring new goods and services.

H) Machines can do better than humans concerning some specific jobs.

Passage Three

Skim the following passage and answer the question 3 in six minutes.

Assistants in record shops are used to receiving "humming queries": a customer comes into the store humming a song he wants, but cannot remember either the title or the artist. Knowledgeable staff are often able to name that tune and make a sale. Hummers, though, can be both off-key and off-track. Frequently, therefore, the cash register stays closed and the customer goes away disappointed. A new piece of software may change this. If Online Music Recognition and Searching (OMRAS) is successful, it will be possible to hum a half-remembered tune into a computer and get a match.

OMRAS, which has just been unveiled at the International Symposium on Music Information Retrieval, in Paris, is the brainchild of a group of researchers from the Universities of London, Indiana and Massachusetts. Music-recognition programs exist already, of course. Mobile-phone users, for instance, can dial into a system called Shazam, hold their phones to a source of music, and then wait for the title and artist to be texted back to them.

Shazam and its cousins work by matching sounds directly to recordings, several million of them, stored in a central database. For Shazam to make a match, though, the music source must be not just similar to, but actually identical with, one of the filed recordings. OMRAS, by contrast, analyses the music. That means it can make a match between different interpretations of the same piece. According to Mark Sandler, the leader of the British side of the project, the program would certainly be able to match performances of the same work by an amateur and a professional pianist. It should also pass the humming-query test.

The musical analysis performed by OMRAS is unlike any that a musicologist would recognize. A tune is first digitized, so that it can be processed. It is then subject to such mathematical indignities as wavelet decomposition, multi-resolution Fourier analysis, polyphase filtering and discrete cosine transformation. The upshot is a mathematical model of the sound that contains the essence of the original, without such distractions as style and quality. That essence can then be compared with a library of known essences and a match made. Unlike Shazam, only one library reference per tune is needed.

So far, Dr Sandler and his colleagues have been restricted to modelling classical music. Their 3,000-strong database includes compositions by Bach, Beethoven and Mozart. Worries about copyright mean that they have not yet gained access to company archives of pop music, though if the companies realize that the consequence of more humming queries being answered is more sales, this may change. On top of that, OMRAS could help to prevent accidental copyright infringements, in which a composer lifts somebody else's work without realizing his inspiration is second-hand. Or, more cynically, it will stop people claiming that any infringement was accidental. There is little point in doing that when a quick check on the Internet could have set your mind at rest that your magnum opus really was yours.

(495 words)

Question 3: The passage is mainly _____.
A) a comparison of two music-recognition programs.
B) an introduction of a new software.
C) a survey of the music recognition and searching market.
D) an analysis of the functions of music recognition softwares.

第二章　寻读查找具体信息

第一节　策略讲解

寻读同略读一样，也是一种快速阅读的技巧。Scan 意为 look through a written passage quickly to find important or interesting information。所谓"寻读（scanning)"，就是通过目光扫视，以最快的速度从一篇文章中披沙拣金，迅速寻找出你所期望得到的某一具体情况、数据等。寻读要利用特殊信息点，比如数字、日期、人名等。以日常生活为例。假如你要买一台价格适中、性能良好的计算机，你可浏览有关计算机广告。各种广告林林总总，品牌不同、功能各异、价格不等。你便可先从价格上考虑，通过寻读，找出几个你准备买的型号，然后从性能、信誉上进行比较，最终选中你想买的那一台。

寻读法也常见于阅读考试的细节题中。寻读的目的主要是有目标地去找出文中某些特定的信息，也就是说，在通过对文章略读以及在信号词的指导下对文章主旨有所了解后，在文中查找与某一问题、某一观点或某一单词有关的信息，寻找解题的可靠依据。这种方法的特点是有的放矢，为我所用。譬如根据提问查找某一人名、地名，某一件事发生的年月或其他类似的情况。寻读时要以很快的速度扫视文章，确定所查询的信息范围。对于阅读理解的细节题，寻读技巧要是使用得当，往往会省时省事，迅速而准确地找出答案。

略读（skimming）和寻读（scanning）的结合使用。首先，在略读时，要花足够的时间去读段落的第一句和第二句，直到完全理解含义，了解作者的写作意图，因为第一句往往是该段的主题句（top sentence），而第二句往往是对前句的延伸或进一步的解释。然后，运用寻读，迅速浏览从第三句开始的后面部分，搜寻作者对开头两句的支持句（supporting sentences），并同时注意文章中间是否有转折词（transitional words），因为这些词常常会把文章的思路逆转或加入其它重要的信息。最后，当读到段落的最后一句时，我们又要使用略读，这时必须再次放慢速度直到完全消化作者对段落的小结（conclusion），因为该小结有可能与主题句截然相反或引导读者进入下一个段落。

以下面这个例子为例，看看如何利用略读和寻读来确定段落的主题句。

Tourism, holiday-making and travel are these days more significant social phenomena than most commentators have considered. On the face of it there could not be a more trivial subject for a book. And indeed since social scientists have had considerable difficulty explaining weightier topics, such as work or politics, it might be thought that they would have great difficulties in accounting for more trivial phenomena such as holiday-making. However, there are interesting parallels with the study of deviance. This involves the investigation of bizarre and idiosyncratic social practices, which happen to be defined as deviant in some societies but not necessarily in others. The assumption is that the investigation of deviance can reveal interesting and significant aspects of "normal" societies. It could be said that a similar analysis can be applied to tourism.

(133 words)

题目：Which one of the following phrases can be the heading of the above paragraph?

选项：A) The politics of tourism.

B) The cost of tourism.

C) Justifying the study of tourism.

D) Tourism contrasted with travel.

E) The essence of modern tourism.

F) Tourism versus leisure.

G) The role of modern tour guides.

【答案与解析】在这篇谈旅游业的段落中，作者第一句和第二句都使用了比较级（more significant than..., there could not be a more trivial...），这时我们要充分理解作者的含义：表面上（on the face of it）是说旅游业不是一个重要的话题，实际上暗示出旅游业的重要性，为下文的反驳作好铺垫。随后的支持句进一步展开，探讨旅游业的困难性，似乎对于旅游业的论述是极其艰难的。然而，文章中的 However 使思路陡然一转，随后把对旅游业的研究与对 deviance 的研究作了一个对比，最后得出的结论是认为研究旅游业是同样有趣和重要的。通过这样的略读和扫描的结合，我们很容易地得出答案为 C) Justifying the study of tourism，其中 justify 意为"为……辩护，证明……有理"。

不管是略读还是寻读，识别"信号词"（signal words）都很关键。"信号词"是指一些在阅读中起着提示作用的词语。这些词语预示着将要读到的内容与上下文存在什么样的关系，或具有什么样的逻辑。因为文章的句子不是无序地排列，而是按照一定关系，有目的、有规律地组织起来的。注意信号词能使我们了解作者的思路，理顺该句与上下文之间的逻辑关系，从而提高阅读理解的效率和准确率。请看下面这一段落：

In that mill, I learned the process of making paper. **First**, the logs are put in the shredder. **Then**, they are cut into small chips and mixed with water and acid. **Next**, they are heated and crushed to a heavy pulp to be cleaned. It is also chemically bleached to whiten it. **After** this, it is passed through rollers to flatten it. **Then**, sheets of wet paper are produced. **Finally**, the water is removed from the sheets which are pressed, dried and refined **until** the finished paper is produced.

分析：作者通过表示先后顺序的信号词 first, then, next, after, then 和 finally，有条不紊地描述出造纸工艺的整个过程。

常用信号词可以归纳为下面几种。

1) 表示递进的信号词：

after all, also, again, and then, moreover, further, further more, additionally, in addition, in other words, moreover, to repeat 等；

2) 预示有相同或类似内容的信号词：

and, also, likewise, in addition, besides, similarly, as well as, the same as 等；

3) 预示有不同或相反内容出现的信号词：

but, however, while, whereas, on the other hand, on the contrary, as apposed to, to the opposite, otherwise 等；

4) 表示因果关系的信号词：

as, for, since, because, as a result, consequently, thus, so, therefore, for this reason, so that, thereby 等；

5) 表示条件性的信号词：

if, in case, assuming that, on condition that, on the supposition that, provided that 等；

6) 表示总结性内容的信号词：

in short, in a word, in brief, briefly, in conclusion, as a result, in sum, to sum up, by and large, to conclude 等；

7) 表示顺序的信号词：

before, after, another, first, next, then, last, finally, afterwards, later on, since then, eventually, in the end, at last 等；

8) 表示解释、举例说明关系的信号词：

for example, for instance, such as, to illustrate, evidently, obviously, in other words, that is to say, the same as 等；

9) 表示目的的信号词：

in order to, in order that, so that, so as to, for the purpose that 等。

第二节 范文阅读

Passage One

Scan the following passage and answer the question in one minute.

Shift key: A key that, when pressed in combination with another key, gives the other key an alternative meaning; for example, producing an uppercase character when a letter key is pressed. The Shift key is also used in various key combinations to create nonstandard characters and to perform special operations. Early IBM keyboards labeled the Shift key only with a white arrow. Later keyboards label this key with the word Shift.

(71 words)

计算机词汇：

uppercase character 大写字符　　　　　　　　letter key 字母键
nonstandard characters 非标准（尺度）字符

Ex. How early IBM keyboards labeled its Shift key?
_____.

【答案与解析】Using a white arrow. 通过 IBM 和 Shift 两个大写单词很快定位到最后一句话。

Passage Two

Scan the following passage and answer the question in one minute.

Portal: A Web site that serves as a gateway to the Internet. A portal which may be a search engine or a directory web page is a collection of links, content, and services designed to guide users to information they are likely to find interesting — news, weather, entertainment, commerce sites, chat rooms, and so on. Yahoo!, Excite, MSN.com , Infoseek, AOL, Lycos and Netscape NetCenter are examples of portals. A web page is the starting point for web surfing.

(80 words)

> 计算机词汇：
>
> gateway *n.* 网关　　　　　　　　　　directory web page 目录网页
>
> portal *n.* 门户网站，入口站点　　　　search engine 搜索引擎

Ex. What kind of information can you find through a portal?

_____.

【答案与解析】Information such as news, weather, entertainment, commerce sites, chat rooms, and so on. 根据关键词"information"及破折号定位第二句。

Passage Three

Scan the following passage and answer the question in one minute.

　　Query By Example (QBE): In database management programs, a query technique, developed by IBM for use in the QBE program, that prompts you to type the search criteria into a template resembling the data record. The advantage of query-by-example retrieval is that you don't need to learn a query language to frame a query. When you start the search, the program presents a screen that lists all the data fields that appear on every data record; enter information that restricts the search to just the specified criteria. The fields left bland, however, will match anything.

(95 words)

> 计算机词汇：
>
> search criteria 搜索条件　　　　　　　　template *n.* (计算机文档的) 模板；样板
>
> retrieval *n.* (计算机系统信息的) 检索；恢复　　query language 查询语言

Ex. What is the advantage of QBE?

_____.

【答案与解析】There is no need to learn a query language to frame a query. 根据关键词"advantage"定位第二句。

Passage Four

Scan the following passage and answer the question in two minutes.

　　Start with early diagnosis. Wearables can detect subtle changes that otherwise go unnoticed, leading to less severe disease and cheaper treatment. Sensors will reveal if an older person's balance is starting to weaken. People's gait (步法) and arm-swing change in early-stage Parkinson's. Strength exercise can help prevent falls and broken limbs. Psychiatric (精神病学的) diagnosis may be enhanced by tracking patterns of smartphone use — without monitoring what people see or type. A smart ring can help a woman conceive, by predicting her menstrual cycle. It can also detect pregnancy less than a week after conception (many women continue to drink or smoke for weeks before they realize they are pregnant).

(108 words)

计算机词汇：

wearable *adj.* & *n.* 可穿戴(的)设备　　　　monitor *v.* & *n.* 监视（器）

Ex. How many examples are used to illustrate the topic?
_____.

【答案与解析】Three. 主旨句为前两句。然后利用传感器检测老年人的平衡能力，智能手机加强精神疾病诊断，智能戒指检测女性怀孕三个例子来证明主旨句。

Passage Five

Scan the following passage and answer the questions in seven minutes.

"The creation of the PC is the best thing that ever happened," said Bill Gates at a conference on "digital dividends" in 2000. He even wondered if it might be possible to make computers for the poor in countries without an electric power grid (电力网络). The answer is yes, and things are going even further. Villagers in a remote region of Laos that has neither electricity nor telephone connections are being wired up to the Internet.

Lee Thorn, the head of the Jhai Foundation, an American-Lao organization, has been working for nearly five years in the Hin Heup district. The foundation has helped villagers build schools, install wells and organize a weaving co-operative. But those villagers told Mr Thorn that what they needed most was access to the Internet. To have any hope of meeting that need, in an environment which is both physically harsh and far removed from technical support, Mr Thorn realized that a strong computer was the first requirement.

He therefore turned to engineers working with the Jhai Foundation, who devised a machine that has no moving, and few delicate parts. Instead of a hard disk, the Jhai PC relies on flash-memory chips to store its data. Its screen is a liquid-crystal display, rather than an energy-guzzling glass cathode-ray tube (阴极射线管) — an exception to the rule that the components used are old-fashioned, and therefore cheap. (No Pentiums, for example, just a 486-type processor.) Mr Thorn estimates that, built in quantity, each Jhai PC would cost around $400. Furthermore, because of its simplicity, a Jhai PC can be powered by a car battery charged with bicycle cranks (自行车曲柄) — thus removing the need for a connection to the grid.

Wireless Internet cards connect each Jhai PC to a solar-powered hilltop relay station which then passes the signals on to a computer in town that is connected to both the Lao phone system (for local calls) and to the Internet. Meanwhile, the Linux-based software that will run the computers is in the final stages of being "localized" into Lao by a group of expatriates in America.

One thing that the new network will allow villagers to do is decide whether it is worth going to market. Phon Hong, the local market town, is 30km away, so it is worth knowing the price of rice before you set off to sell some there. Links further afield may allow decisions about growing crops for foreign markets to be taken more sensibly — and help with bargaining when these are sold. And there is also the pleasure of using Internet telephony to talk to relatives who have gone to the capital, Vientiane, or even abroad.

If it works, the Jhai PC and its associated network could be a widespread success. So far, the foundation has had expressions of interest from groups working in Peru, Chile and South Africa. The prototype (原型模型) should be operational in Laos this December and it, or something very much like it, may soon be bridging the digital divide elsewhere as well.

(495 words)

计算机词汇：

moving parts 活动部件　　　　　hard disk 硬盘　　flash-memory chip 闪存芯片
liquid-crystal display 液晶显示器　Internet telephony 网络电话

Ex.1. From the second paragraph, we learn that_____.

　　A) villagers are isolated from the outside world.

　　B) Lee's work is to improve the life of people living in the countryside.

　　C) the harsh environment keeps Lee from doing better job.

　　D) engineers have moved to far-away towns due to the poverty of the villages.

Ex.2. Which of the following is NOT a feature of Jhai PC?

　　A) Delicateness.　　B) Practicality.　　C) Simplicity.　　D) Low cost.

Ex.3. Which of the following statements is true according to the text?

　　A) The Jhai PC has no expensive parts.

B) The Jhai PC is powered by solar energy.
C) The project has gained support from non-resident Laotians.
D) The software that runs the Jhai PC is a local product.

【答案与解析】Ex.1 B，属事实细节题。根据文章第二段，Lee 在该地区工作五年来为该地区作了很多工作，比如建学校、打井、组织合作社等。这些工作无疑是为了改善人们的生活。

Ex.2 A，属事实细节题。从文章第二段最后一句话"Mr. Thorn realized that a robust computer was the first requirement"可以看出该项目需要的是一种非常"robust"的电脑。robust 的本意是"结实的，强健的"，在此转义为"耐用的，不易坏的"。文章第三段提到这种电脑时说它没有活动部件，"delicate"的部件也很少。"delicate"的意思除了"精致的"以外，还有"脆弱的"意思，可见这台电脑是属于结实耐用不易坏的那种。

Ex.3 C，属事实细节题。文章第四段最后一行提到，该电脑的运行软件正在由一些移居美国的老挝人进行本地化的工作。可见这一项目得到了非常住居民的支持。

第三节　阅读演练

Passage One

Scan the following passage and answer the question 1 in one minute.

These concerns have substance. Virtual work risks entrenching silos (隔开；孤立): people are more likely to spend time with colleagues they already know. Deep relationships are harder to form with a laggy (延迟的) internet connection. A study from 2010 found that physical proximity between co-authors was a good predictor of the impact of scientific papers: the greater the distance between them, the less likely they were to be cited. Even evangelists for remote work make time for physical gatherings.

(77 words)

计算机词汇：
virtual work 线上办公　　　remote work 远程工作　　　physical gatherings 线下聚会

Question 1: Is the sentence below TRUE or FALSE based on the passage?
(　　) According to the study in 2010, there is little impact of the scientific papers written by co-authors.

Passage Two

Scan the following passage and answer the question 2 in one minute.

Bored with common greetings that may go unnoticed? WeChat is offering an alternative to saying "Hi." In the latest version, users can double-click any user's avatar and WeChat will notify the other party you want to talk to. Users can also nudge (轻触) themselves, though this serves no purpose beyond alleviating boredom. The nudge function also works on group chats, and any nudge to a group will be visible to all members. That being said, you may want to think twice before you nudge your boss: who knows whether the reply will be a smile or being booted (踢) from the group.

(102 words)

计算机词汇：

double-click 双击　　　　　　　group chats 群聊

avatar *n.*（在电脑游戏或网络中代表使用者的）图标，头像

Question 2: What is the function of nudge?

_____.

Passage Three

Scan the following passage and answer the question 3 in one minute.

As he renames Facebook as Meta, Mark Zuckerberg is placing a bold bet on his vision of the future: a techno-utopia (科技乌托邦) where people are connected and live in a virtual universe. Zuckerberg said, "It is an iconic social media brand, but increasingly it just doesn't encompass everything that we do." MARK Zuckerberg wants to turn Facebook into metaverse, a three-dimensional world where users can socialize, do shopping and play games. Meta, speculates Mr Muzellec, could sponsor initiatives to reduce teen online addiction while Facebook carries on hooking (吸引) the kids. Meta may have just half the letters in Facebook's name, but it retains all of the company's problems.

(109 words)

计算机词汇：

virtual universe 虚拟世界　　　　three-dimensional *adj.* 三维的，立体的

> **Metaverse** *n.* 虚拟实境；3D 虚幻世界（尤指角色扮演游戏创造的世界）；元宇宙（虚拟空间）

Question 3: What can you do with Meta?
_____.

Passage Four

Scan the following passage and answer the question 4 in one minute.

  Query Language: In database management programs, a retrieval and data-editing language you use to specify what information to retrieve and how to arrange the retrieved information on-screen or when printing. The dot-prompt language of dBASE is a full-fledged query language, as is Structured Query Language (SQL), which is used for minicomputer and mainframe databases and is growing in popularity in the world of personal computers. The ideal query language is a natural language, such as English.

(76 words)

> 计算机词汇：
> query language 查询语言 retrieval *n.*（计算机系统信息的）检索；恢复
> data-editing 数据编辑 dot-prompt 点提示符 mainframe database 大型机数据库

Question 4: Is the sentence below TRUE or FALSE based on the passage?
( ) English is a good retrieval and data-editing language.

Passage Five

Scan the following passage and answer the questions 5-9 in ten minutes.

  DOTCOMS may be moribund, but inside companies, the Internet is still finding cost-saving new uses. "B2E — business-to-employee — didn't have a crash," says Bipin Patel, in charge of developing the potential of the corporate intranet at the Ford Motor Company. "It's still growing."
  Ford has gone further than most companies to get its employees online: it offered its American employees personal computers, and 90% of them accepted. Ford hopes that the free PCs will save its own and its employees' time by moving services online. General Motors, Ford's

great rival, considered a similar scheme but found that most employees willing to use PCs already had them. It is helping staff to pay for high-bandwidth connections instead.

At Ford, the human-resources department has pioneered a scheme to provide up-to-the-minute information to employees about pay and benefits. In the past, employees sometimes found that it took weeks to get a copy of the pay information they needed to do their tax returns, and the department's staff spent mind-numbing hours answering the same questions from hundreds of different employees. Now employees can look at a password-protected site that displays their payslips (工资单) over the previous 18 months. They can see all deductions, and the hours they worked. All this information was on the human-resources database: displaying it to employees has saved staff time.

"People want more and more of this self-service information, which they can manage themselves," observes Mr Patel. "There is no such thing as information overload here, because it's their information." Even training seems to work better online: Ford employees can now download a range of courses, including one on "Listening and Handling Tough Situations", all designed for digestion in 10 or 15 minutes. The company claims to have cut training costs by $2m during the past six months, as fewer people leave their desks to learn.

The company also uses its intranet to communicate with its staff around the world. Jacques Nasser, Ford's boss, sends out "Let's chat" notes once a week. In fact, Mr Nasser does most of the chatting. He gets hundreds of e-mails in reply, but the communication is basically a one-way flow. The company also runs chat-rooms, in which employees can question various in-house experts and outside analysts live on the corporate intranet.

In time, thinks Mr Patel, communications technology will reshape corporate behaviour. It will encourage collaboration and team-working. Already, the Internet is causing disintermediation within companies, he argues, just as it did in e-commerce: the human-resources department does much less administration once the benefits system is more self-service, but rather more advising and consulting. One day, working in human resources might even be fun.

(440 words)

计算机词汇：

intranet *n.* 内联网；企业内部网　　　bandwidth *n.* [电子][物] 带宽；[通信] 频带宽度
password *n.* 密码；口令
B2E　即"企业对职员"，它强调的是职员，而不是客户或者企业。B2E 解决日益突出的 IT 人才短缺状况。从某种程度上来说，它使企业调动一切可能的因素来吸引和挽留高素质的 IT 人才，比如：大胆的创新理念、优厚的待遇、深造机会、灵活的上班时间、丰厚的奖金以及职员授权策略等。
intranet　内联网，企业网的一种，是采用因特网技术组建的企业网。在内联网中除了

> 提供 FTP、E-mail 等功能外，还提供 Web 服务。
> disintermediation　　脱媒现象，即进行交易时跳过所有中间人而直接在供需双方间进行。

Question 5: How does the author introduce his topic?

　　A) Posing a contrast.　　　　　　　C) Explaining a phenomenon.

　　B) Justifying an assumption.　　　　D) Making a comparison.

Question 6: Why did Ford and GM intend to provide their employees with PCs?

　　A) PCs can help them save a lot.

　　B) B2E is a growing thing.

　　C) The employees prefer to use PCs.

　　D) Proving PCs is a way of competition.

Question 7: Which of the following is not the advantage of internet?

　　A) It can save time.

　　B) It can save cost.

　　C) It encourages collaboration and team-working.

　　D) It makes human-resources department an easy job.

Question 8: The expression "a one-way flow" (Line 3, Paragraph 5) most probably means _____.

　　A) not encouraging open answers.　　C) only yes or no questions.

　　B) only one side asking questions.　　D) the topics lack variety.

Question 9: Which of the following is true according to the text?

　　A) The Internet can help DOTCOMS come back to life.

　　B) The courses downloaded are practical, but time-consuming.

　　C) The internet makes the work of human-resources department more direct and interesting.

　　D) The employees can manage all the information by themselves.

Unit Two

第二单元

阅读技巧：理解文章结构

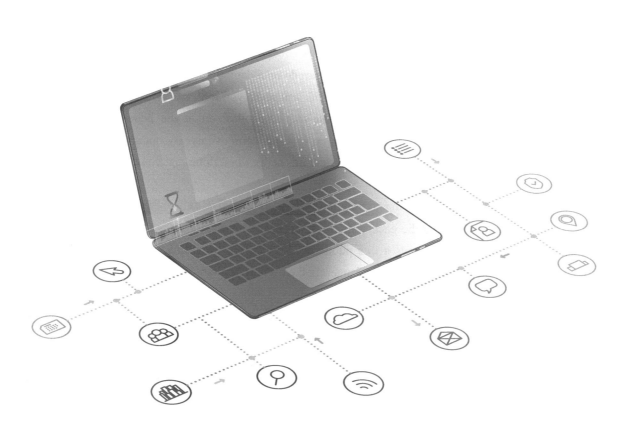

文章结构（或者文本结构）指的是信息在文章中的组织方式。在一部作品中，甚至在一个段落中，文本的结构都可能发生多次变化。文本结构通常有七种常用的组织模式。

1. 因果关系 (Cause and Effect): 解释某事的结果。

E.g. The dodo bird used to roam in large flocks across America. Interestingly, the dodo wasn't startled by gun shot. Because of this, frontiersmen would kill entire flocks in one sitting. Unable to sustain these attacks, the dodo was hunted to extinction.

2. 时间顺序 (Chronological): 文章中的信息是按照时间顺序排列的。

E.g. Jack and Jill ran up the hill to fetch a pail of water. Jack fell down and broke his crown and Jill came tumbling after.

3. 比较和对比 (Compare and Contrast): 描述两个或两个以上的事物，讨论了它们的异同。

E.g. Linux and Window are both operating systems. Computers use them to run programs. Linux is totally free and open source, so users can improve or otherwise modify the source code. Windows is proprietary, so it costs money to use and users are prohibited from altering the source code.

4. 重要性顺序 (Order of Importance): 信息按层次或优先级表示。

E.g. Here are the three worst thing that you can do on a date. First, you could tell jokes that aren't funny and laugh really hard to yourself. This will make you look bad. Worse though, you could offend your date. One bad "joke" may cause your date to lash out at you, hence ruining the engagement. But the worst thing that you can do is to appear slovenly. By not showering and properly grooming, you may repulse your date, and this is the worst thing that you can do.

5. 问题和解决方案 (Problem and Solution): 描述问题，并提出或解释响应或解决方案。

E.g. Thousand of people die each year in car accidents involving drugs or alcohol. Lives could be saved if our town adopts a free public taxi service. By providing such a service, we could prevent intoxicated drivers from endangering themselves or others.

6. 顺序/过程写作 (Sequence/Process Writing): 信息是按步骤组织的，或者一个过程是按发生的顺序解释的。

E.g. Eating cereal is easy. First, get out your materials. Next, pour your cereal in the bowl, add milk, and enjoy.

7. 空间/描述性写作 (Spatial/Descriptive Writing): 信息按照空间顺序(从上到下，从左到右)组织。

E.g. When you walk into my bedroom there is a window facing you. To the right of that is a dresser and television and on the other side of the window is my bed.

  英语文章，一般讲究思想脉络分明，论理清晰明确。英语文章有这样的特点关键在于段落是构建在一个以"主题"(topic)为核心的基础上。全段以此为中心，进行逻辑发展，使得内容"单一连贯"。阅读文章时，须通过分辨主题句、找关键词等捕捉文章的主题意义，并分析总结段落的各个信息如何为文章的主题服务，使文章的主题更明了。作者运用逻辑关系使其所有的思想逻辑清晰地互相关联。一般说来，一个段落中的细节，1)可以依其重要性，由轻而重安排；2)可以依其时间发生的顺序，由先而后陈述；3)可以依其空间的位置，顺次交代；4)可以用演绎的方式，由结论而至细节逐一证明；5)可以以归纳的方法，由诸多细节的陈述，而得出统一的结论。另一方面，了解细节的逻辑安排方法，有助于我们了解段落内句与句之间的关系，了解段落之间的关系，这样在文章主题思想的指导下，就可以很容易地追踪到作者的思想脉络，从而迅速有效地判断各段落间的内容，理清思路，抓住重点。

  再从文体（记叙文、描写文、说明文和议论文）的基本特征角度讲解文章行文结构。1)记叙文事件的叙述连贯并符合逻辑，通常按照时间顺序展开。事件之间通常使用一些过渡性词语来连接。文中细节逐渐引向事件发生、发展的高潮部分。2)描写文强调的是人的感受（人的视觉、听觉、触觉、味觉和嗅觉等几大感官感受）。人物描写通过他或她所说的或所做的反映人的性格、思想和感情；地点描写应该按照一定的顺序展开，由远到近或由近到远，由左至右或由右至左，由上到下或由下到上；景物描写通常包括三个主要部分：背景、人物和行为。这些能够给我们提供一定线索，快速捕获答案。3)说明文阐述事物的原理、起因和可能的后果。说明文的行文中常出现下列规律：举例说明，分析过程，比较和对比，分类以及因果分析等。说明文的基本模块是：引言部分，文章主体，结束段。结束段的行文多以下列方式展现：用立意新颖的句式重述主题，或总结文章的要点，或者重述加总结。4)议论文是人们交流思想的一种重要形式，其最终目的是说服听众或读者接受某个观点。议论文要用逻辑和论据来影响别人的看法或行动，而不是仅仅依靠动之以情，据理力争。议论文的行文规则为：提出论点，充分、恰当地提出论据并论证论点，得出结论。

第一章 明确主要支撑细节

第一节 策略讲解

  主旨并不能提供全部信息，事实和细节出现在阅读的段落中，有助于发展段落的主要思想。这些事实和细节可以描绘出一幅更完整的画面，可以提供例子来帮助更好地理解观点，证明一个观点，展示这个观点是如何与其他观点联系起来的。

一段话中并非所有的事实都具有同样的重要性。然而，提供主要思想信息的细节是非常重要的。一些小细节或不太重要的细节有助于充实段落。不过，如果我们的目标是清晰快速地理解所读内容，我们可能会忽略小细节。因此，在寻找段落的主要细节时，我们首先要学会寻找段落的大意。当我们找到所有有助于构成这段话主旨的事实和细节后，我们就可以把主要细节和次要细节分开了。

当阅读一段文字时，我们应该首先试着识别出主要细节，然后试着把主要细节从小细节或不太重要的细节中分离出来。以下有一些方法可以帮助我们找到主要细节。

1) 抓住大意阅读。如果能很容易地找出主要观点，那么支持该观点的事实就会很突出。

2) 事实和细节并非同等重要。只寻找与主旨有关的事实。

具体步骤为：①用略读(skimming)的方法通读材料，大概了解原文，掌握其主旨。②按文章的体裁、作者写作的组织模式及有关信息词，如 for example、first、second 等预测应该到何处寻找所需要的事实。③把主要精力放在寻找所需要的细节上。快速通篇跳读，眼睛自左至右、自上而下呈 Z 形扫视，直到找到所需要的部分。待找到所需要的部分时，可放慢速度，细读要查找的内容。

在平时阅读训练中，应学会快速辨认和记忆事实或细节，带着问题寻找答案。略读阅读材料，将注意力集中在与 who、what、when、where 问题有关的细节或数字信息。对某些细节，可一面阅读一面概括归纳，尽力记住这些主要细节，并留意它们所在位置。然后浏览材料及复读阅读材料，复读时通篇阅读，寻找与问题有关的细节。最后解答问题，确定答案。

第二节　范文阅读

Passage One

Read the following passage carefully and answer the questions.

Valeta Young, 81, a retiree from Lodi, Calif., suffers from congestive (充血的) heart failure and requires almost constant monitoring. But she doesn't have to drive anywhere to get it. Twice a day she steps onto a special electronic scale (秤), answers a few yes or no questions via push buttons on a small attached monitor and presses a button that sends the information to a nurse's station in San Antonio, Texas. "It's almost a direct link to my doctor," says Young, who describes herself as computer illiterate but says she has no problems using the equipment.

Young is not the only patient who is dealing with her doctor from a distance. Remote monitoring is a rapidly growing field in medical technology, with more than 25 firms competing to measure remotely — and transmit by phone, Internet or through the airwaves — everything from patients' heart rates to how often they cough.

Prompted both by the rise in health-care costs and the increasing computerization of health-care equipment, doctors are using remote monitoring to track a widening variety of chronic diseases. In March, St. Francis University in Pittsburgh, Pa., partnered with a company called BodyMedia on a study in which rural diabetes patients use wireless glucose (葡萄糖) meters and armband sensors to monitor their disease. And last fall, Yahoo began offering subscribers the ability to chart their asthma (哮喘) conditions online, using a PDA-size respiratory (呼吸的) monitor that measures lung functions in real time and e-mails the data directly to doctors.

Such home monitoring, says Dr. George Dailey, a physician at the Scripps Clinic in San Diego, "could someday replace less productive ways that patients track changes in their heart rate, blood sugar, lipid levels, kidney functions and even vision."

Dr. Timothy Moore, executive vice president of Alere Medical, which produces the smart scales that Young and more than 10,000 other patients are using, says that almost any vital sign could, in theory, be monitored from home. But, he warns, that might not always make good medical sense. He advises against performing electrocardiograms (心电图) remotely, for example, and although he acknowledges that remote monitoring of blood-sugar levels and diabetic ulcers (溃疡) on the skin may have real value, he points out that there are no truly independent studies that establish the value of home testing for diabetes or asthma.

Such studies are needed because the technology is still in its infancy and medical experts are divided about its value. But on one thing they all agree: you should never rely on any remote testing system without clearing it with your doctor.

(425 words)

计算机词汇：

armband sensors 臂带传感器 computer illiterate. 电脑盲
computerization n. 计算机化，电脑化 digitized computerization. 数字化电脑处理
PDA Personal Digital Assistant 个人数字助理（又称掌上电脑）

Ex.1. How does Young monitor her health conditions?

    A) By stepping on an electronic scale.

    B) By answering a few yes or no questions.

    C) By using remote monitoring service.

D) By establishing a direct link to her doctor.

Ex.2. Which of the following is not used in remote monitoring?
   A) Car.   B) Telephone.   C) Internet.   D) The airwaves.

Ex.3. Why is Dr. Timothy Moore against performing electrocardiograms remotely?
   A) Because it is a less productive way of monitoring.
   B) Because it doesn't make good medical sense.
   C) Because it's value has not been proved by scientific study.
   D) Because it is not allowed by doctors.

Ex.4. Which of the following is true according to the text?
   A) Computer illiterate is advised not to use remote monitoring.
   B) The development of remote monitoring market is rather sluggish.
   C) Remote monitoring is mainly used to track chronic diseases.
   D) Medical experts agree on the value of remote monitoring.

【答案与解析】本文是一篇说明文，介绍了远程监护目前的发展状况、它的优势以及对它的相反意见等。第一段以一位病人通过远程监护就医的例子引入话题。第二段介绍了远程监护市场的情况。第三段介绍了远程监护在医疗保健领域的应用。第四段引用医生的话说明这种监护方式的优点。第五段介绍反对的声音。最后一段得出结论：在没有得到医生的许可下，绝不能依赖任何远程测试系统。

Ex.1. C，属事实细节题。文中第一段讲 Young 在家中每天上两回电子称，通过电子秤上的一台小型监视器上的按钮回答一些答案为"是"或"否"的问题，然后再按一个按钮把信息送到得克萨斯州圣安东尼奥的一个护士台。第二段开头说，Young 并不是远程就医的唯一病人，可见她是通过远程监护服务来监测自己的健康情况的。

Ex.2. A，属事实细节题。第一段提到 Young 的身体监护时说，"she doesn't have to drive anywhere to get it"。第二段提到各种测量数据是通过 "phone, Internet or through the airwaves"传送的，可见汽车是不用于远程监护的。

Ex.3. B，属事实细节题。文中第五段提到 Dr. Timothy Moore 时说，"他警告说，那种做法并不总是 make good medical sense。" 接下来就列举他反对 performing electrocardiograms remotely 的例子。可见做远程心电图的医学价值不大。

Ex.4. C，属事实细节题。文章第三段第一句提到医生们正在利用远程监护监测 a widening variety of chronic diseases。可见远程监护主要是用于监测慢性疾病。这个答案也可以从文中提到的应用远程监测的充血性心力衰竭、糖尿病、哮喘等疾病推断出来。

Passage Two

There is a passage with ten statements. Each statement contains information given in one of the paragraphs. Identify the paragraph from which the information is derived. You may choose a paragraph more than once. Each paragraph is marked a letter.

Pessimism vs Progress

A) Faster, cheaper, better-technology is one field many people rely upon to offer a vision of a brighter future. But as the 2020s dawn, optimism is in short supply. The new technologies that dominated the past decade seem to be making things worse. Social media were supposed to bring people together. Today they are better known for invading privacy, spreading propaganda and undermining democracy. E-commerce, ride-hailing and the gig economy may be convenient, but they are charged with underpaying workers, exacerbating inequality and clogging the streets with vehicles. Parents worry that smartphones have turned their children into screen-addicted zombies.

B) The technologies expected to dominate the new decade also seem to cast a dark shadow. Artificial intelligence (AI) may well entrench bias and prejudice, threaten your job and shore up authoritarian rulers (see Essay in this issue). Autonomous cars still do not work, but manage to kill people all the same. Polls show that internet firms are now less trusted than the banking industry. At the very moment banks are striving to rebrand themselves as tech firms, internet giants have become the new banks, morphing from talent magnets to pariahs. Even their employees are in revolt.

C) The New York Times sums up the encroaching gloom. "A mood of pessimism", it writes, has displaced "the idea of inevitable progress born in the scientific and industrial revolutions." Except those words are from an article published in 1979. Back then the paper fretted that the anxiety was "fed by growing doubts about society's ability to rein in the seemingly runaway forces of technology".

D) Today's gloomy mood is centered on smartphones and social media, which took off a decade ago. Yet concerns that humanity has taken a technological wrong turn, or that particular technologies might be doing more harm than good, have arisen before. In the 1970s the despondency was prompted by concerns about overpopulation, environmental damage and the prospect of nuclear immolation. The 1920s witnessed a backlash against cars, which had earlier been seen as a miraculous answer to the affliction of horse — drawn vehicles — which filled the streets with noise and dung, and caused congestion and accidents. And the blight of industrialization was decried in the 19th century by Luddites, Romantics and socialists, who worried (with good reason) about the displacement of skilled artisans, the despoiling of the countryside and the suffering of factory hands toiling in smoke-belching mills.

E) Stand back, and in each of these historical cases disappointment arose from a mix of unrealized hopes and unforeseen consequences. Technology unleashes the forces of creative destruction, so it is only natural that it leads to anxiety; for any given technology its drawbacks sometimes seem to outweigh its benefits. When this happens with several technologies at once, as today, the result is a wider sense of techno-pessimism.

F) However, that pessimism can be overdone. Too often people focus on the drawbacks of a new technology while taking its benefits for granted. Worries about screen time should be weighed against the much more substantial benefits of ubiquitous communication and the instant access to information and entertainment that smartphones make possible. A further danger is that Luddite efforts to avoid the short-term costs associated with a new technology will end up denying access to its long-term benefits — something Carl Benedikt Frey, an Oxford academic, calls a "technology trap". Fears that robots will steal people's jobs may prompt politicians to tax them, for example, to discourage their use. Yet in the long run countries that wish to maintain their standard of living as their workforce ages and shrinks will need more robots, not fewer.

G) That points to another lesson, which is that the remedy to technology-related problems very often involves more technology. Airbags and other improvements in safety features, for example, mean that in America deaths in car accidents per billion miles travelled have fallen from around 240 in the 1920s to around 12 today. AI is being applied as part of the effort to stem the flow of extremist material on social media. The ultimate example is climate change. It is hard to imagine any solution that does not depend in part on innovations in clean energy, carbon capture and energy storage.

H) The most important lesson is about technology itself. Any powerful technology can be used for good or ill. The internet spreads understanding, but it is also where videos of people being beheaded go viral. Biotechnology can raise crop yields and cure diseases — but it could equally lead to deadly weapons.

I) Technology itself has no agency: it is the choices people make about it that shape the world. Thus the techlash is a necessary step in the adoption of important new technologies. At its best, it helps frame how society comes to terms with innovations and imposes rules and policies that limit their destructive potential (seat belts, catalytic converters and traffic regulations), accommodate change (universal schooling as a response to industrialization) or strike a trade-off (between the convenience of ride-hailing and the protection of gig-workers). Healthy skepticism means that these questions are settled by a broad debate, not by a coterie of technologists.

Fire up the moral engine

J) Perhaps the real source of anxiety is not technology itself, but growing doubt about the ability of societies to hold this debate, and come up with good answers. In that sense, techno-pessimism is a symptom of political pessimism. Yet there is something perversely reassuring about this: a gloomy debate is much better than no debate at all. And history still argues, on the whole, for optimism. The technological transformation since the Industrial Revolution has helped curb ancient evils, from child mortality to hunger and ignorance. Yet, the planet is warming and antibiotic resistance is spreading. But the solution to such problems calls for the deployment of more technology, not less. So as the decade turns, put aside the gloom for a moment. To be alive in the tech-obsessed 2020s is to be among the luckiest people who have ever lived.

计算机词汇：

e-commerce *n.* 电子商务　　　　　ride-hailing 打车服务

gig economy 零工经济，指的是区别于传统"朝九晚五"、时间短且灵活的工作形式，利用互联网和移动技术快速匹配供需方。由日渐崛起的由 90 后、00 后等新世代主导的互联网和移动互联网，正在颠覆一些传统的经济形态。"零工经济"作为一种新的经济形态在悄然流行，近年更呈现爆发式的增长。

Luddite 勒德分子，十九世纪初英国手工业工人中参加捣毁机器的人。在现代用法中，勒德分子这个词用以描述工业化、自动化、数位化或新技术的抵制者。举例来说，如果多数人使用手机或沉迷网络，通常就会出现另一群勒德分子(Luddite)，拒用手机或 PDA，以保护隐私为理由拒绝上网，坚决地向科技说不。

autonomous car 自动驾驶汽车　　　　　techno-pessimism 技术悲观主义

techlash 科技抵制潮，《经济学人》中出现的合成词，指对科技巨头如"谷歌""脸书"的抵制。科技抵制潮是基于科技公司日益增长的影响力，而生成的一种强烈和广泛的负面情绪。

Ex.1. People pay too much attention to the disadvantages of a new technology but take its advantages for granted.

Ex.2. There is a lesson that more technology is needed to tackle the technology-related problems.

Ex.3. Whether a new important technology should be adopted should be settled by a broad debate.

Ex.4. Long ago people worried that some technologies might be more harmful.

Ex.5. Any powerful technology itself can do good or harm.

Ex.6. The Internet companies has turned into the new banks while the banks are trying to become tech company.

Ex.7. Those who live in the 2020s obsessed with technology is the luckiest.

Ex.8. To avoid the short-term costs caused by a new technology will fall into a technology trap.

Ex.9. It seems that new technology is bring problems after the year 2020.

Ex.10. Pessimism has taken place of the idea of necessary progress brought by the scientific and industrial revolution.

【答案】Ex.1. F　Ex.2. G　Ex.3. I　Ex.4. D　Ex.5. H　Ex.6. B　Ex.7. J　Ex.8. F　Ex.9. A　Ex.10. C

第三节 阅读演练

Passage One

Read the following passage carefully and answer the question 1.

  The computer virus is an outcome of the computer overgrowth in the 1980s. The cause of the term "computer virus" is the likeness between the biological virus and the evil program infected with computers. The origin of this term came from an American science fiction *The Adolescence of P-1* written by Thomas J. Ryan, published in 1977. Human viruses invade a living cell and turn it into a factory for manufacturing viruses. However, computer viruses are small programs. They replicate by attaching a copy of themselves to another program.

  Once attached to one host program, the viruses then look for other programs to "infect". In this way, the virus can spread quickly throughout a hard disk or an entire organization when it infects a LAN or a multi-user system. At some point, determined by how the virus was programmed the virus attacks. The timing of the attack can be linked to a number of situations, including a certain time or date, the presence of a particular file, the security privilege level of the user, and the number of times a file is used. Likewise, the mode of attack varies. So-called "benign" viruses might simply display a message, like the one that infected IBM's main computer system last Christmas with a season's greeting. Malignant viruses are designed to damage the system. The attack is to wipe out data, to delete flies, or to format the hard disk.

  What kind of viruses are there? There are four main types of viruses: shell, intrusive, operating system and source code. Shell viruses wrap themselves around a host program and don't modify the original program. Shell programs are easy to write, which is why about half of viruses are of this type. Intrusive viruses invade an existing program and actually insert a portion of themselves into the host program. Intrusive viruses are hard to write and very difficult to remove without damaging the host file.

计算机词汇：

computer virus 电脑病毒 evil program 有害程序 replicate *n.* 复制
copy *n.* 副本 host program 主程序 infect *v.* 感染（计算机病毒）
LAM（local area network）局域网 multi-user system 多用户系统
file *n.* （计算机的）文档 benign virus 良性病毒 malignant virus 恶性病毒
format *v.* 格式化 shell *n.* (short for "shell program") 外壳程序
source code 源代码 host program 主程序

Question1: Are the statements below TRUE or FALSE? Please write "T" for True and "F" for False in corresponding brackets.

1. (   ) The computer viruses are just like human viruses.
2. (   ) The viruses will infect the programs once entering the computer.
3. (   ) After the virus infects a local area network, it can spread throughout a hard disk quickly.
4. (   ) The virus attack follows a fixed mode.
5. (   ) The benign viruses will format the hard disk.
6. (   ) Comparing with shell programs, intrusive viruses are harder to write.

Passage Two

Read the following passage carefully and choose the best answer to the question 2.

That little "a" with a circle curling around it that is found in email addresses is most commonly referred to as the "at" symbol. Surprisingly though, there is no official, universal name for this sign. There are dozens of strange terms to describe the @ symbol.

Before it became the standard symbol for electronic mail, the @ symbol was used to represent the cost or weight of something. For instance, if you purchased 6 apples, you might write it as 6 apples @ $1.10 each.

With the introduction of e-mail came the popularity of the @ symbol. The @ symbol or the "at sign" separates a person's online user name from his mail server address, for instance, webmaster@chinaacc.com. Its widespread use on the Internet made it necessary to put this symbol on keyboards in other countries that have never seen or used the symbol before. As a result, there is really no official name for this symbol.

The actual origin of the @ symbol remains an enigma (谜).

History tells us that the @ symbol stemmed from the tired hands of the medieval monks. During the Middle Ages before the invention of printing presses, every letter of a word had to be painstakingly transcribed by hand for each copy of a published book. The monks that performed these long, tedious copying duties looked for ways to reduce the number of individual strokes per word for common words. Although the word "at" is quite short to begin with, it was a common enough word in texts and documents that medieval monks thought it would be quicker and easier to shorten the word "at" even more. As a result, the monks looped the "t" around the "a" and created it into a circle — eliminating two strokes of the pen.

计算机词汇：

electronic mail 电子邮件   server *n.* 服务器

Question 2: Which of the following is NOT represented by @?

    A) The weight and cost of something.

    B) The A in the keyboard.

    C) The separation between user's name and his server address.

    D) The word "at" in texts and documents.

Passage Three

Read the following passage carefully and choose the best answer to the question 3.

    In Chinese, a computer is popularly known as an "electrical brain", for the working process of a computer is similar to a human brain very much.

    Just as a driver can't drive a car without driving skills or the car itself, you can't control a computer without controlling techniques or the computer itself. The controlling techniques are called software, while computers themselves and related devices are called hardware. The work of a computer is just making full use of various resources by software set in the computer, and directing the hardware to realize marvelous omnipotent (无所不能的) functions.

    There are many types of microcomputers. Here, we will use an IBM Personal Computer (PC) to illustrate the primary components of a microcomputer. Other brands and models of microcomputers exhibit difference in appearance and operations. An IBM PC's primary hardware components are the main frame, the monitor, the keyboard, and many peripherals such as the disk drive, hard disk, printer, and mouse, all of which are hardwired to the main frame. The main frame is the heart of a microcomputer system. It contains the Central Processing Unit (CPU), a chip that controls the major operations of the computer and the main memory.

计算机词汇：

microcomputer 微型计算机　　model 型号　　operation 操作

main frame 主机　　monitor 监视器　　keyboard 键盘

peripheral 外围设备　　disk drive 磁盘驱动器　　hard disk 硬盘

mouse 鼠标　　CPU(Central Processing Unit) 中央处理单元　　chip 芯片

main memory 主存储器

Question 3: Which of the following is NOT the hardware component of an IBM PC?

    A) Keyboard.　　　　B) Printer.　　　　C) CPU.　　　　D) Mouse.

第二章　区分事实和观点

第一节　策略讲解

阅读时我们应该思考作者是在展示一个既定的事实还是个人观点。事实一旦经过核实或来自有信誉的来源，就可以被接受并被视为可靠的信息。然而，意见并不是可靠的信息来源，应该受到质疑和仔细评估。可以把文章中的内容区分是观点态度抑或是事实是一个阅读者应该具备的基本技能。在文章里若有一些类似表示观点态度的词(如 maintain、argue)出现，那么这些词后面出现的应该就是观点和态度。观点和态度是主观的，不能被证明。但若是如此的表达：evidence show、experiment suggest 等，后面跟的就可能是事实，可以被证明。

1. 定义事实

一般来说，事实是实际发生过的事情，或者是经验上正确的，可以有证据支持的事情。它们可以在官方文件、法律记录、科学研究成果、特定技术数据或其他可靠来源中找到。事实通常用精确的数字或数量、重量和度量以及客观的语言来表达。

2. 确定意见

观点是对个人感觉或判断的陈述，它反映的是一种解释，而不是证据的积累。观点通常以事实为基础，但它们也包括作者对事实的个人解释，这可能与读者的解释不一致。即使是仔细研究同一问题的专家们，对这个研究问题的看法也往往非常不同。尽管观点的准确性无法得到验证，但作者应该用证据、事实和理由来支持他们的观点——用任何能支持观点的信息，让读者相信这是一个有效的观点。有些作者在表达观点时，会小心地向读者发出信号。下面是一些表达观点的常用词汇。

1) 赞同/积极：positive, favorable, approval, enthusiasm, supportive, defensive, concerned, confident, interested, optimistic, impressive;

2) 否定/消极：negative, disapproval, objection, opposition, critical, criticism, disgust, contempt, compromise, worried, indifferent, depressed, pessimistic, unconcerned, hostile;

3) 怀疑：suspicion, suspicious, doubt, doubtful, question, puzzling;

4) 客观：objective, neutral, impartial, disinterested, unprejudiced, unbiased, detached;

5) 主观：subjective, indifference, tolerance, pessimism, gloomy, optimistic, sensitive, scared, reserved, consent, radical, moderate, mild, ironic, confused, amazed, worried, concerned, apprehensive, biased, indignant.

有时候作者会把事实和观点混在一起，以赢得读者对某个特定观点的认同。说服技巧包括引用发表观点的消息来源，用"it is a fact"来套用，然后再加上一个伪装的观点。例如，"The author believes that George Washington was the first president of the United States."是一个事实，而"It is a fact that George Washington was the best president of the United States."只是一个观点。再比如，"They are putting too much carbon dioxide into the atmosphere, which prevents heat from escaping from the earth into space."是一个事实，而 "The earth may become too hot for the lives to live on."是一个观点。

尝试区分以下10个句子是事实还是观点：

A) It is snowing outside.

B) It is a nice day.

C) Daisies are the prettiest of all flowers.

D) Tiger Woods was the first African American to win the Masters Golf Tournament.

E) A pair of women stockings is made from one strand of nylon that is 4 miles long and knitted into 3 million loops.

F) Originality and independence are the most difficult qualities to achieve.

G) I believe it is harder to divide than multiple numbers.

H) Fridays come before Saturdays on the calendar.

I) He feels spinach tastes better than candy.

J) Teachers should allow students to use calculators during tests.

【答案与解析】

A) 事实，因为它可以在个人观察下证明真假。

B) 它是一种表达个人感觉或判断的观点。例如，如果下雪，孩子们可能会认为这是一个美好的一天，而街道清洁工认为这是一个糟糕的一天。像"nice"这样的形容词通常表示一种观点。

C) 它是一种观点，因为它只是表达了个人的感觉或判断。最高级"the prettiest"表示一种观点。

D) 这是有历史文献支持的事实。

E) 这是有具体技术数据支持的事实。

F) 它是一种表达个人感觉或判断的观点。最高级"the most difficult"表示一种观点。

G) 它是一种表达个人感觉或判断的观点。"believe"和比较级"harder"表示观点。

H) 这是一个有证据支持的事实。

I) 它是一种表达个人感觉或判断的观点。"feel"和比较级"better"表示一种观点。

J) 它是一种表达个人感觉或判断的观点。"should"通常表示一种观点。

第二节 范文阅读

Passage One

Read the following passage and decide which sentences are facts and which are opinions. Write the number of the sentences in the corresponding space.

Some people say that ① parents should plan their children's leisure time carefully. Take me for example. ② My parents were very strict with me during my childhood. ③ My leisure time was divided into several parts: reading, writing, playing and so on. However, others think that children should decide for themselves how to spend their free time. The result is that ④ some children don't spend their free time in a planned way. They play just as they like without any plan at all. So far as I am concerned, I think the question is not whether parents should plan their children's leisure time carefully, but whether they can do that properly. ⑤ Some parents don't understand their children while some parents spoil their children too much. So it is not an easy thing for the parents to help shape their child to be a true man. ⑥ But I think, with love and understanding, nothing can be hard.

Ex.Facts: _____

Ex.Opinions: _____

【答案】Facts: ①, ⑥

　　　　Opinions: ②, ③, ④, ⑤

【答案与解析】句①是作者引用别人的观点，带有主观性，所以它是观点。句②、句③是作者拿自己举例，是用事实来证明观点，所以属于事实。句④是为了指出第二段开头的观点的弊端而举的反例，属于事实。句⑤提到的内容是客观存在的，所以是事实。句⑥是作者提出的自己的看法，带有主观性，属于观点。

Passage Two

Read the following passage and answer the question.

Just as crying can be healthy, not crying — holding back tears of anger, pain or suffering — can be bad for physical health. Studies have shown that too much control of emotions can lead to high blood pressure, heart problems and some other illnesses. If you have a health problem, doctors will certainly not ask you to cry. But when you feel like crying, don't fight it. It's a natural — and healthy — emotional response.

Ex.: According to the author, which of the following statements is true?

  A) Crying is the best way to get help from others.

  B) Fighting back tears may cause some health problems.

  C) We will never know our deep feelings unless we cry.

  D) We must cry if we want to reduce pressure.

【答案与解析】B。作者在文章中说 Studies show that too much control can lead to high blood press, heart problems and other illnesses，即过多地忍住情感（如忍住不哭等）有可能会导致健康问题。这里虽然用了 Studies show… (研究表明……)这样的字眼，但作者在此显然是为增加说服力而特意采用的一种表现手法，也就是说，研究所表明的结果就是作者的观点，故最佳答案为 B。特别说明：阅读理解中的推断题通常涉及的是作者的看法、意图与态度，即作者本人在文章字里行间所表达的观点或看法。

Passage Three

Read the following passage and answer the question after.

  What is the most popular sport in the world? If you said "soccer", you are right. People of all ages play it. Some people even play it in their bare feet. They use rolled-up rags for a ball. A soccer ball weighs one pound. The players like to use a ball with stripes or a pattern. It makes it easy to see the ball spin during a game. A ball can reach a speed of 75 miles per hour. In the United States we call the game soccer. The rest of the world calls it football. When soccer first came to the United States, we already had a sport called football. The word soccer comes from England. Soccer is the best name for the game. How many times can you bounce a soccer ball off your body in 30 seconds? You may use your head, knee, or foot. You cannot use your hands. The ball must not hit the ground. Ferdie Adoboe holds the record. He did it 141 times! The biggest soccer ball ever made is in a mall in the Middle East. It is 13 feet, one inch tall. It has 32 pieces and was sewn by hand. The biggest soccer ball must be an amazing sight.

Ex.: Are these statements in the above passages Fact or Opinion? Write "F" for fact, "O" for opinion in the following blankets.

  1) _____ People of all ages play soccer.

  2) _____ Soccer balls look great with stripes or patterns.

  3) _____ Soccer balls can go 75 miles per hour.

  4) _____ The United States is the only place that calls the game soccer.

5) _____ It is very hard to bounce a soccer ball for 30 seconds.
6) _____ Ferdie Adoboe bounced the ball 141 times in 30 seconds.
7) _____ The biggest soccer ball is 13 feet, one inch tall.
8) _____ The biggest soccer ball is an amazing sight.

【答案】1) F 2) O 3) F 4) F 5) O 6) F 7) F 8) O

第三节　阅读演练

Passage One

Read the following paragraph and answer the question 1.

More and more businesses work closely with companies in other countries. They need many different kinds of workers who can communicate in different languages and understand other cultures. No matter what career you choose, if you've learned a second language, you'll have a real advantage. A technician who knows Russian or German, the head of a company who knows Japanese or Spanish, or a salesperson who knows French or Chinese can work successfully with many more people and in many more places than someone who knows only one language.

Professionals who know other languages are called on to travel and exchange information with people in other countries throughout their careers. Knowing more than one language enhances opportunities in government, business, medicine and health care, law enforcement, teaching, technology, the military, communications, industry, social service, and marketing. An employer will see you as a bridge to new clients or customers if you know a second language.

Question 1: Are the underlined statements in the above passages Fact or Opinion? Write "F" for fact, "O" for opinion in the following blankets.

1) _____ More and more businesses work closely with companies in other countries.
2) _____ No matter what career you choose, if you've learned a second language, you'll have a real advantage.
3) _____ An employer will see you as a bridge to new clients or customers if you know a second language.

Passage Two

Read the following passage and answer the question 2.

A study by a team of Purdue University researchers suggests that even moderate exercise may lead to reduced iron in the blood of women. The study of 62 formerly inactive women who began exercising three times a week for six months was published in the journal Medicine & Science in Sports & Exercise.

Exercise can result in iron loss through a variety of mechanisms. Some iron is lost in sweat, and for unknown reasons, intense endurance exercise is sometimes associated with bleeding of the digestive system. Athletes in high-impact sports such as running may also lose iron through a phenomenon where small blood vessels in the feet leak blood.

If iron levels are low, talk with a physician to see if the deficiency should be corrected by modifying your diet or by taking supplements. In general, it's better to undo the problem by adding more iron-rich foods to the diet, because iron supplements can have serious shortcomings.

Eat iron-rich foods with plentiful Vitamin C. On the other hand, people who are likely to have low iron should avoid drinking coffee or tea with meals.

Question 2: Are the underlined statements in the above passages Fact or Opinion? Write "F" for fact, "O" for opinion in the following blankets.

1) _____ The study of 62 formerly inactive women who began exercising three times a week for six months was published in the journal Medicine & Science in Sports & Exercise.
2) _____ Some iron is lost in sweat, and for unknown reasons, intense endurance exercise is sometimes associated with bleeding of the digestive system.
3) _____ If iron levels are low, talk with a physician to see if the deficiency should be corrected by modifying your diet or by taking supplements.
4) _____ On the other hand, people who are likely to have low iron should avoid drinking coffee or tea with meals.

Passage Three

Read the following passage and answer the question 3.

Not all countries and classes are adopting online dating at the same rate or in the same way. Americans are charging ahead; Germans, comparatively, lagging behind. India, which has long had a complex offline market for arranged marriages within religious and caste boundaries, has seen it move online. Last year saw a rare Indian tech-sector IPO when matrimony.com raised 500 crore rupees ($70m) to help it target the marriage market.

In countries where marriage is still very much in the hands of parents, today's apps offer an option which used hardly to exist: casual dating. Yu Wang, the chief executive of <u>Tantan, founded in 2015 and now one of China's largest dating apps</u>, says the country's offline dating culture is practically non-existent. "If you approach someone you don't know and start flirting, you're a scoundrel," he says. But on Tantan "you don't expose yourself, there's no danger of getting rejected, you cannot lose face." As of February, Tantan had 20m users and had created some 10m couples, <u>Mr Wang says, adding: "That's a significant effect on society."</u>

How much happiness these particular possibilities for granularity have brought about is not known. But there are some figures for the field as a whole. In a 2013 study researchers from Harvard University and the University of Chicago showed that marriages that started online were less likely to end in break-up and were associated with higher levels of satisfaction than marriages of the same vintage between similar couples who had met offline: <u>the difference was not huge, but it was statistically significant. Couples who met online also reported being slightly more satisfied with their marriage than those who met offline, by an average of one fifth of a point more on a seven-point scale.</u> Scaled up to the third or more of marriages in America that start online, that would mean that close to a million people have found happier marriages than they would have otherwise thanks to the internet — as have millions more around the world.

Question 3: Are the underlined statements in the above passages Fact or Opinion? Write "F" for fact, "O" for opinion in the following blankets.

1) _____ Not all countries and classes are adopting online dating at the same rate or in the same way.
2) _____ Tantan, founded in 2015 and now is one of China's largest dating apps.
3) _____ Mr Wang says, adding: "That's a significant effect on society."
4) _____ The difference was not huge, but it was statistically significant.
5) _____ Couples who met online also reported being slightly more satisfied with their marriage than those who met offline, by an average of one fifth of a point more on a seven-point scale.

Unit Three

第三单元
阅读技巧：明确作者的意图和态度

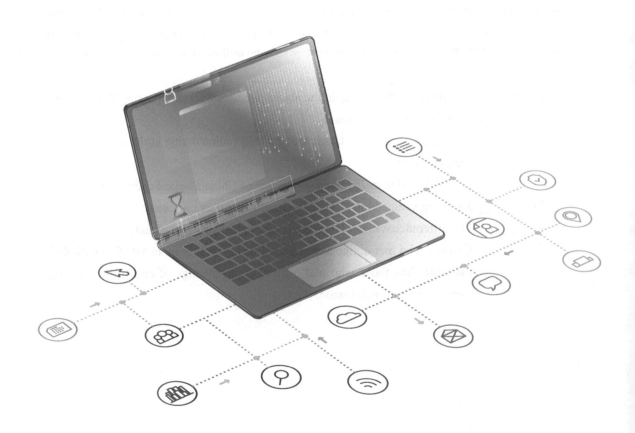

第一章　明确作者的写作意图

第一节　策略讲解

一个作者写作有很多原因，写作意图（writing purpose）就是作者写一篇文章的原因。询问作者写作意图类的问题就是问作者为什么要写这篇文章。这类问题有多种形式，如："Why does the author write this?" "What is the reason this passage is written for?"等，或者题干中常有 purpose 或表示目的的动词不定式 intend to、mean to、in order to 等。

作者写作主要有五类目的：

1）说服（argue, convince, prove）；
2）叙述（relate a sequence of events, tell a story）；
3）描述（appeal to the senses, create a picture or impression）；
4）娱乐（amuse, divert attention from reality）；
5）告知（state, teach, explain）。

作者运用某种写作手法的目的或是要引出主题、突出主题，或是要吸引读者关注主题，或是要把主题说明得更清楚。无论哪种手法，也不管出于何种目的，都必定与主题相关，为主题服务。

推断作者的写作意图有以下两个方法，可综合考虑：

1. 主旨推断法。

写作意图与文章主旨密切相关，因此，明确作者的写作意图和确定文章的主旨大意题一样，采用略读法，即重点关注文章首尾段和各段的首尾句，找到主题句，抓住文章主旨，然后由主旨来推断作者的写作目的。议论文、说明文、新闻报道等的主题句多在文首。

2. 文体推断法。

作者的写作目的与文体密切相关。

1）议论文的目的通常是说服读者接受或赞同某一观点，倡导某种做法等。
2）说明文的目的是使读者获得某种知识，提出某种建议、劝告或呼吁，或希望引起有关部门或人士对某现象给予重视。
3）记叙文的目的一般是分享一段有趣的经历，告诉读者一个有趣的故事，使读者获得乐趣；若是夹叙夹议的文章，则是表达作者的感悟或给读者某种教育或启示。
4）广告是作者要推销一种产品或一种服务，因此其目的是吸引更多人。

例如：

Direction: Please read the following passages, ask yourself "Why does the author write

this?", and then match each passage with the author's purpose listed after.

Passage One: NVIDIA GeForce 8 uses an unprecedented design and Unified Shader architecture to deliver robust performance. Full hardware support for DirectX10 features such as Geometry Shaders, stream out, improved instancing and Shader Model4.0. Support for these features makes the Geforce8800 GPU extremely high-performance. All DirectX9, OpenGL, and previous DirectX programs and games have a high performance on the Unified Shader design of the GeForce 8 GPU. The cost is $421. Order yours today!

Passage Two: Video Wall is a wall of multiple rack-mounted video monitors or rear projection displays that together produce a very large image or combinations of images. Video walls can be dramatic attention-grabbing devices for advertising or art when a single image is spread across many adjacent monitors. Video walls can be fed from computer output or from real time television signals. Software divides the image and routes the pieces to each monitor, and can create special effects.

Passage Three: Artificial Life is a scientific research area devoted to the creation and study of computer simulations of living organisms. Computer viruses have forced a renewal of the debate on the definition of life. Besides forcing us to re-examine our definition of life, artificial life research may create more effective technology. By applying artificial life concepts to real problems, we can program computer-generated solutions to compete for survival based on their capability to perform a desired task well.

Question: The author write this to:

  A) promote.

  B) persuade/convince.

  C) teach/inform.

【答案与解析】

Passage One：选 A。本文先是介绍了 NVIDIA GeForce 8 的特点，最后号召读者 "Order yours now!"，意味着作者想把产品卖给读者。

Passage Two：选 C。本文向读者介绍了视频墙的特点和功能，目的是让读者了解该产品。

Passage Three：选 B。本文介绍了人工生命研究的给人类带来的好处，目的是劝服或鼓励读者支持人工生命的研究。

第二节　范文阅读

Passage One

Read the following passage and decide what the author's purpose is in writing this passage.

  Nonetheless, the impact of these firms on business will get bigger in two ways. First, they will grow fast — although whether fast enough to justify their valuations remains to be seen. To maximize their chances, many are offering not just a single service (such as search or video), as Western firms tended to in their early years, but a bundle of services in one app instead, in the hope of making more money per user. This approach was pioneered in China by Alibaba and Tencent. Go-Jek in Indonesia offers ride-hailing, payments, drug prescriptions and massages. Facebook is pushing a digital payments system in India through its chat service, WhatsApp.

The second is that in the emerging world, established firms are likely to be disrupted more quickly than incumbents (在职者) were in the rich world. They have fewer infrastructures, such as warehouses and retail sites, to act as a barrier to entry. Many people, especially outside the big cities, lack access to their services entirely. Beer, shampoo and other consumer-goods firms could find that as marketing goes digital, new insurgent brands gain traction faster. Banks will be forced to adapt quickly to digital payments or die. Viewed this way, there is a huge amount of money at stake — the total market value of incumbent firms in the emerging world, outside China, is $8trn. If you thought the first half of the internet revolution was disruptive, just wait until you see the second act.

(245 words)

计算机词汇：

ride-hailing 网约车   digital payment 电子支付

Ex. The author writes this passage in order to_____.
  A) inform.　　　　B) instruct.　　　　C) predict.　　　　D) persuade.

【答案与解析】B. 文中作者介绍了互联网公司扩大其影响力的两种方式，并对这两种方式造成的影响或结果做出评价，给予读者一定的启示，引发读者对于互联网是否扰乱了市场的思考。

Passage Two

Read the following passage and decide what the author's purpose is in writing this passage.

  Should video websites have to review content before they publish it? Where does the boundary lie between hate speech and incitement to violence? Is pornography created by artificial intelligence an invasion of privacy? These are all hard questions, but behind them lies an even more difficult one: who should provide the answers?

On the internet, such dilemmas are increasingly being resolved by private firms. Social networks are deciding what kinds of misinformation to ban. Web-hosting companies are taking down sites they deem harmful. Now financial firms are more actively restricting what people can buy.

The digital gatekeepers are doing a mixed job. But it is becoming clear that it ought not to be their job at all. The trade-offs around what can be said, done and bought online urgently need the input of elected representatives. So far governments have been better at complaining than at taking responsibility.

(147 words)

计算机词汇：

social network 社交网络　　web-hosting company 网络托管公司
digital gatekeeper 数字守门人

Ex. The author writes this passage in order to_____.

　　A) inform.　　　　B) instruct.　　　　C) predict.　　　　D) persuade.

【答案与解析】D. 作者在第一段提出了网络应该由谁来监管的问题，第二段指出这类问题目前多是由私人公司来完成，第三段号召民众和政府参与网络监管。

Passage Three

Read the following passage and decide what the author's purpose is in writing this passage.

　　The human face is a remarkable piece of work. The astonishing variety of facial features helps people recognize each other and is crucial to the formation of complex societies. So is the face's ability to send emotional signals, whether through an involuntary blush or the artifice of a false smile. People spend much of their waking lives, in the office and the courtroom as well as the bar and the bedroom, reading faces, for signs of attraction, hostility, trust and deceit. They also spend plenty of time trying to dissimulate.

　　Technology is rapidly catching up with the human ability to read faces. In America facial recognition is used by churches to track worshippers' attendance; in Britain, by retailers to spot past shoplifters. This year Welsh police used it to arrest a suspect outside a football game. In China it verifies the identities of ride-hailing drivers, permits tourists to enter attractions and lets people pay for things with a smile. Apple's new iPhone use it to unlock the homescreen.

(171 words)

计算机词汇：

facial recognition 人脸识别   homescreen   *n.* 主屏幕；待机界面

Ex. The author writes this passage in order to_____.
   A) inform.   B) instruct.   C) predict.   D) persuade.

【答案与解析】 A. 文中作者只是客观地描述事实，告知读者人脸识别技术的应用。作者并未发表任何个人观点。

第三节　阅读演练

Passage One

Read the following passage and answer the question 1.

　　Japan is getting tough about recycling — and not in the paper and plastic kind of way. Recently, the country requires that all electronic goods — TVs, VCRs, stereos, and more — be recycled. But recycling will not be left to consumers, instead, the devices will be sent to the original manufacturer for proper disposal.

　　The new law poses a few challenges to manufacturers who are now rushing to set up collection networks and perfecting techniques to disassemble and recycle older products. With an eye toward the future, they are also integrating easily recycled materials into new products. Plastics, a major component of most electronic products, pose a particular obstacle because their quality becomes worse and worse with age, losing strength and flexibility even if reprocessed. NEC Corp. overcomes this problem by creating a plastics sandwich, in which the filling is 100 percent recycled plastic and the outer layers a mixture of 14 percent recycled material. The resulting plastic has sufficient strength and toughness for use as a case for desktop PCs. The company, in cooperation with plastic maker Sumitomo Dow, has also developed a new plastic, which engineers claim retains its mechanical properties through repeated recycling. NEC uses the plastic, which is also flame-retardant (阻燃的) in battery cases for notebook PCs.

Meanwhile, Matsushita Electric, maker of the Panasonic brand, is avoiding plastic in favor of magnesium (镁). Magnesium, says the company, is ideal for recycling because it retains its original strength through repeated reprocessing. Matsushita has developed molding techniques to form magnesium into the case for a 21-inch TV. Unfortunately, the magnesium case and energy-saving features make the TV about twice as expensive as an ordinary model. The company hopes, however, that increased use of magnesium will eventually bring prices down.

(299 words)

Question 1: The author writes this passage in order to_____.
    A) inform.   B) instruct.   C) predict.   D) persuade.

Passage Two

Read the following passage and answer the question 2.

There are subtle, complex changes taking place in human communication, thought, and relationships within online communication and information communities. The Web is part of these changes, enabling new forms of communication and information delivery, and brings about new associations among people. One challenge for our society is to find a solution to the questions raised by these changes. How might our culture, society, and communication patterns change as a result of widespread Web use?

The Web can link information in useful ways, giving rise to new insights — transformation of information to knowledge. The Web offers immediate delivery of information to specialized audiences. Before the invention of computer networks, an individual could not easily seek out several hundred others interested in a specialized hobby or area of interest, when those people were spread worldwide. No traditional media offered a personally available means to accomplish this. But the Web does fill this "media gap" and this feature is certainly a contributor to the Web's popularity and growth.

Associative linking promotes relationships among people in addition to relationships among information. Experts in a particular field create pools of knowledge on their home page, and other people link to these pages, groups of experts form. These groups might be based on information or on hobbies, interests, culture, or political leanings. The result is that "electronic tribe" can form that gather people in associations that could not be possible any other way.

As the Web alters communication and information patterns, the resulting change raises issues our society must face for individual, group, and societal responsibility. Moral and legal issues will arise in the areas of individual behavior, societal responsibility for issues of access and information literacy, and the new relationships, communications, and thought patterns the Web promotes.

(293 words)

计算机词汇：

associative linking 联想链接。在计算机科学中，指两个或多个数据元素之间的关系，这些关系可用于搜索、排序和其他操作。

electronic tribe 电子部落

information literacy 信息素养；资讯素养。资讯素养是一种"使人能够更有效地寻找、选择及评估传统或网上资源的技巧"。是确认资讯、检索及寻获资讯、组织及整理资讯、使用及创造资讯、评估资讯的能力。

Question 2. The author writes this passage in order to_____.

A) present what challenges the computers bring.

B) prove that the Web can replace all other means of communication.

C) show what changes the Web bring and how we should face them.

D) show that the Web can fulfill our needs of knowledge.

Passage Three

Read the following passage and choose the best answer to the questions 3-7.

In a ditty for the stage, W.S. Gilbert once gave warning that "Things are seldom what they seem/Skim milk masquerades as cream." If appearances were tricky in 1878, they have just become trickier still. By doubling the resolution of existing liquid-crystal displays (LCDS), IBM has created a monitor which, when viewed from 18 inches away or farther, shows images that the human eye finds indistinguishable from the real thing.

The T220, as it is called, measures 22 inches across the diagonal (对角线), and displays 9.2m picture elements ("pixels"). That gives it a resolution of 200 pixels per inch, twice the previous state of the art. This achievement has come as a result of gradual improvements in optics (光学), liquid-crystal chemistry and microelectronics made by IBM groups in Yamato, Japan, and Yorktown Heights, New York.

LCDs work by sandwiching a thin sheet of liquid crystals — in this case, thin-film transistors — between two narrowly separated panes of glass. Typically, small glass spheres have held the two panes of glass apart, damaging by refraction the performance of the display. IBM has replaced the spheres with small posts, which are located in the interstices (裂缝) between pixels, and they do not disturb the light as it leaves the excited liquid crystal. In the past, attempts to achieve such high pixel rates have been stymied by the build-up of electrical static, which caused problems with the brightness of the screens. The IBM groups have solved this by using a laser to scan back and forth across the glass, preventing the build-up of static electricity.

At a current retail price of $22,000, the T220 is hardly going to be flying off the shelves. But it will be ideal for hospitals. Historically, radiology (放射学) has been a driving force behind the development of high-resolution screens. And the T220's price tag will go almost unnoticed when attached to MRI (magnetic resonance imaging) or CT (computerized tomography) scanning machines. Until now, no monitor has been able to display the 5m pixels of data that a typical CT-scanning machine produces. The ability to reproduce the data with perfect fidelity should help radiologists make more accurate diagnoses from the computer screen.

According to Bob Artemenko, director of marketing and strategy for IBM'S business display unit, the new screen could also help petroleum engineers to speed up their analysis of where to drill from one month to one day. Similarly, the higher fidelity will allow CAD (computer-aided design) systems, especially in the motor and aerospace industries, to work faster — because the detail revealed by the new monitor can cut out costly prototype-building exercises. RAM'S idea is that the new monitor will allow designers of all sorts to go straight from computer image to final product, eliminating many costly and time-consuming middle stages.

With prices of more conventional 15 inch LCDs now below $500, IBM is expected to shift its engineering effort from achieving high resolution to lowering costs. How long before the T220 starts showing up in high-end laptops? Judging from previous experience, it could happen sooner than most people think.

(522 words)

计算机词汇：

liquid-crystal display 液晶显示器
pixel *n.* 像素
high-resolution screen 高分辨率屏幕
high-end laptop 高端笔记本电脑
monitor *n.* 显示器
thin-film transistor 薄膜晶体管
computer-aided design (CAD) 电脑辅助设计

Question 3. How does the author introduce the topic?

A) Posing a contrast. C) Making a comparison.

B) Justifying an assumption. D) Explaining a phenomenon.

Question 4. Which of the following is not the advantage of the T220?

A) Reasonable price. B) Time-saving. C) Cost-saving. D) High fidelity.

Question 5. The expression "stymied" (Line 6, Paragraph 3) most probably means _____.
    A) limited.   B) controlled.   C) improved.   D) hindered.

Question 6. Why does the hospital ignore the price tag of T220?
    A) T220 creates a driving force for the medical staff.
    B) T220 guarantees a more accurate diagnosis.
    C) CT-scanning machine fails to produce such high fidelity images.
    D) T220 owns the ability to reproduce data with perfect fidelity.

Question 7. What is the current problem IBM facing?
    A) Achieving high resolution.   C) Lowering the price.
    B) Pursuing more applications.   D) Reducing the size of the screen.

第二章 明确作者的观点态度

第一节 策略讲解

一篇文章不可避免地反映了作者的观点、态度和情绪。能否正确把握作者的观点和态度也是体现阅读能力的重要方面。一般来说，对于作者总的态度和倾向，必须在通读全文、掌握了主题思想和主要事实后，方能做出判断。在判断作者观点态度时，我们应注意以下几个问题。

一、有时作者先介绍了某一种观点，却接着在后面提出了相反的观点，因此，要正确判断作者的态度或观点，必须将上下文联系起来看，要注意文章中所陈述的内容并非都代表了作者的观点。

例 1:

Direction: Decide whether or not each of the following statements approves of the book written by Professor Baker.

1) Professor Baker's publisher has stated that this new book will soon take the place of all the old standard works in this field; in view, however, of both the style and content of Professor Baker's book, I find this claim most difficult to accept.

2) When I first opened the package containing a copy of Professor Baker's latest book and read its title, I must admit I felt a sudden sinking of the heart; yet once I had gathered courage to begin my reading, I found the work so far beyond my wildest hopes that I actually missed supper

rather than put the volume down unfinished.

【答案与解析】在 1)段中，作者在前面引用了出版者的赞扬之辞，而后接着表明难以接受此说法，说明他对此书持否定态度。

在 2)段中，作者先说刚看到此书书名时很失望，然后用"yet"表示转折，以至到最后爱不释手，表明他赞许此书。

例 2：

Direction: Decide which of the following statements was written by a person in favor of small-town life.

1) Possibly there are those who derive pleasure from turning back the clock and seeking out the virtues we fondly imagine to have been associated with life in small towns in bygone days. Yet the most superficial reading of contemporary accounts dealing with such an existence makes it quite clear that full enjoyment of life's true pleasures would be much more appreciated.

2) It is true that city-dwellers and suburbanites can enjoy certain facilities that may be denied to the inhabitants of small towns. What they miss, however, far outweighs such advantages, which, in any case, the vast majority rarely has the time or energy to take advantage of.

【答案与解析】在 1)文中，作者在谈到喜欢小城镇生活的人时用了 Possibly there are...(可能有这样的人)来表示不肯定的语气，在谈到小城镇生活的长处时用了...we imagine to have been associated with...(据我们想象……与相联系)来减弱其客观性，暗示未必是真是如此。可见该作者不赞同小城镇的生活。

在 2)文中，作者在谈到城市生活的方便时用了 certain 一词以表示其有限性，并用 maybe denied to the inhabitants of small towns (小城镇的居民也许没有)来表示不肯定。然而在谈到城市生活所失去的东西时用了 far outweighs such advantages (大大超过了这些好处)来加以强调。可见这篇文章的作者是赞同小城镇生活的。

二、关注作者的措辞，看作者在文中用了什么样的口气。若用褒义词，显然是赞成；若用贬义词，显然是反对；若是客观陈述，则是中性的立场，不偏不倚。

作者态度题中经常出的表征态度的形容词如下：

A) 支持或赞成：positive, approving, supportive, optimistic, sympathetic, complimentary, affectionate, confident, appreciative, similar, identical.

B) 中立或客观：neutral, objective, impartial, unbiased, detached.

C) 怀疑、批评或反对：negative, disapproving, critical, pessimistic, doubtful, questioning, suspicious, skeptical, scornful, contemptible, opposite, cynical.

例如：

Direction: Read the following paragraph and answer the question.

Yet there are limits to what a society can spend in this pursuit. As a physician, I know the most costly and dramatic measures may be ineffective and painful. I also know that people in

Japan and Sweden, countries that spend far less on medical care, have achieved longer, healthier lives than we have. As a nation, we may be over-funding the quest for unlikely cures while under-funding research on humbler therapies that could improve people's lives.

Question: In contrast to the U.S., Japan and Sweden are funding their medical care_____ .

    A) more flexibly.　　　　　　　　C) more cautiously.

    B) more extravagantly.　　　　　　D) more reasonably.

【答案与解析】作者用 limits、ineffective、painful 这些消极的词语暗示了美国医疗卫生系统的缺憾，通过把握这些词语，我们就能得出 D 为正确选项。

三、通过关注作者的举例角度和讲解角度，来判断作者的态度倾向：如果作者一直论述某事物积极向上的方面，其态度基本上是积极乐观的；如果作者举例论证某观点时，给的例子是正面的，那么同样可以判断作者的态度是积极乐观的；如果作者的论述有好有坏，举例有正面有反面，可以判断作者的态度是客观的。

例如：

Direction: Read the following paragraph and answer the question after.

Under the new Northern Territory law, an adult patient can request death — probably by a deadly injection or pill — to put an end to suffering. The patient must be diagnosed as terminally ill by two doctors. After a "cooling off" period of seven days, the patient can sign a certificate of request. After 48 hours the wish for death can be met. For Lloyd Nickson, a 54-year-old Darwin resident suffering from lung cancer, the NT Rights of Terminally Ill law means he can get on with living without the haunting fear of his suffering: a terrifying death from his breathing condition. "I'm not afraid of dying from a spiritual point of view, but what I was afraid of was how I'd go, because I've watched people die in the hospital fighting for oxygen and clawing at their masks," he says.

Question: The author's attitude towards euthanasia (安乐死) seems to be that of _____.

    A) opposition.　　　B) suspicion.　　　C) approval.　　　D) indifference.

【答案与解析】作者用一个肺癌病人为例，该病人认为安乐死法案的通过意味着自己可以平静地度过最后的时光，不用担心临死前要遭受的折磨。很明显这个事例是认可安乐死法案的通过是一件好事，属于正面的例子，故正确答案为 C。

第二节　范文阅读

Passage One

Read the following passage and answer the question.

If you're in a job performing tedious tasks, you might think that a robot could do the work instead. But perhaps we underestimate how much technology already helps with the activities that we would otherwise have to do. And as artificial intelligence progresses, we might find it replaces us in the workplace altogether. For now, robotic technology is providing a helping hand for businesses, particularly in manufacturing, assisting humans in performing work more efficiently and sometimes more accurately. For online shopping, for example, robots have become essential in giant warehouses. They sort and move millions of objects of all different shapes and sizes, although humans are still needed to pick and distribute the goods.

The advancement of robotics in the workplace is good for some businesses: the ones who research, develop, build and use them. The British government estimates that by 2035, artificial intelligence could add around £630bn to the UK economy. But there are still tasks that robots can't yet do, and that's the challenge for companies such as Automata. Its co-founder, Suryansh Chandra, told the BBC that his technology will eliminate boring, repetitive jobs that humans don't like and aren't very good at, and also create new ones that are likely to replace them.

It seems inevitable that robots will eventually be able to do more and more of the jobs that are currently performed by humans, so should we be worried by the rise of the machines? Some experts fear hundreds of thousands of jobs could disappear as robots replace human workers. A report by the OECD suggests that 14% of jobs are "at high risk of automation" and 32% of jobs could be "radically transformed", with the manufacturing sector at the highest risk. But as complete automation is some way off, for now we'll have to work side-by-side with our robot colleagues and manage to get along with them before they learn to kick us out of the door!

(323 words)

计算机词汇：
robot *n.* 机器人　　　robotic *adj.* (像)机器人的；机械的
robotics / roʊˈbɑːtɪks / *n.* 机器人技术　　automation *n.* 自动化
automata *n.* 自动装置；小机器人（automaton 的复数）

Ex. What is the author's attitude towards artificial intelligence?
　　A) Worried.　　　B) Suspicious.　　　C) Objective.　　　D) Optimistic.

【答案与解析】C. 作者指出随着人工智能的发展，必然会由人工智能做越来越多由人类完成的工作，建议在完全的自动化到来之前，人类应该学会同"机器人同事"和谐相处。作者的态度中立客观。

Passage Two

Read the following passage and answer the question.

Virtual high schools, which allow students to take classes via PC, have emerged as an increasingly popular education alternative, particularly for on-the-go athletes. University of Miami Online High School (UMOHS), a virtual school that caters to athletes, has more than 400 students enrolled, 65% of whom are athletes. Accredited by the 100-year-old Southern Association of Colleges and Schools, UMOHS offers honors and advanced-placement classes. All course material is online, along with assignments and due dates. For help, says principal Howard Liebman, "a student may e-mail, instant message or call the teacher."

Educators are split on the merits of such schools. Paul Orehovec, an enrollment officer for the University of Miami, admits, "I was somewhat of a skeptic. But when I looked into their programs and accreditation, I was excited. UMOHS is the first online school to be granted membership in the National Honor Society." Kevin Roy, Elite's director of education, sees pitfalls and potential in virtual schools. "You will never have that wonderful teacher who inspires you for life," says Roy. "But the virtual school offers endless possibilities. I don't know where education's imagination will take this."

Ex. To which of the following is the author likely to agree?

A) The education provided by virtual schools is yet to be recognized by authorities.

B) Educators are divided as to whether students should take virtual schools.

C) Despite the defects, virtual schools show great potentials.

D) Regular schools will be replaced by virtual schools sooner or later.

【答案与解析】C。本文介绍了网上虚拟学校的发展状况。文章第二段说一位教务处长既看到了虚拟学校的不足，也看到了它们的潜力。并引用他的话说："虚拟学校却提供了无限可能性。"可见虚拟学校虽然有缺陷，但仍然有着巨大的潜力。

Passage Three

Read the following passage and answer the question.

Muffin Man has more than 2,000 songs on his hard drive, and he's happy to share them. He's a big fan of bands like Pearl Jam and the White Stripes, so there's plenty of hard rock in his collection.

But chances are you'll never get to it. The 21-year-old pizza cook, who asked to be identified by his online nickname, makes his songs available only through private file-sharing networks known as darknets. Unlike such public networks as Kazaa or Morpheus, which let you share songs with anyone, private networks operate more like underground nightclubs or secret societies. To gain access, you need to know the name of the group and a password. And the only way to get that information is from another member who invites you in. Some darknets even encrypt files and mask your identity within a group to keep eavesdroppers (偷听者) from finding out who you are and what you are sharing.

It's a handy invention now that the recording industry has taken to suing kids who share music online. But darknets are not just for digital music files. Carving out a bit of privacy online has wide appeal; students, community groups and even political dissidents can use these hidden networks to share projects, papers and information. One part of the allure is anonymity; the other is exclusivity. Since participation is limited, file searches don't turn up a lot of junk or pornography. Darknets offer the convenience of the Web without a lot of the bad stuff.

You need special software to start a darknet of your own. The two most popular programs are Direct Connect by NeoModus (at neomodus.com) and an open-source variation of it called DC++, available at sourceforge.net. More than 800,000 copies of DC++ have been downloaded since mid-July. A third program, called Waste (also at sourceforge.net), automatically encrypts files but is much harder to use.

There are no good estimates of how many people use darknets. Lowtec, a college sophomore studying computer engineering, figures that 10% of the students at his school (which he declined to name) share files through Direct Connect. "It's much faster than Kazaa," he says. That's because private networks typically link small, close-knit communities in which all members have superfast connections.

The recording industry so far hasn't put much effort into combatting the secret networks, but its neglect might not last long. If networks like Kazaa become too risky, darknets could quickly rise to take their place. And if that happens, the music industry could find itself chasing users who are that much harder to catch.

(430 words)

计算机词汇：

hard drive 硬盘　　　　online nickname 网名　　　　file-sharing network 文件共享网络

encrypt /in'kript/ vt. 加密

Ex. From the text we can see that the writer seems_____.

　　A) positive.　　　　B) negative.　　　　C) doubtful.　　　　D) uncertain.

【答案与解析】A. 本篇是一篇说明文，介绍了一种名为"地下网络"的私人文件共享网络。

整篇文章基本上没有选择负面的材料，而且第三段在介绍 darknets 为何吸引人的时候，作者还讲到了这种私人网络"既提供了网络的便利，又摆脱了网络的糟粕"的优势。在结尾，作者说如果可以共享音乐的公共网络倒下，这种 darknets 就会迅速崛起并取而代之。可见作者对这种网络所持的是一种肯定的态度。

第三节　阅读演练

Passage One

Read the following passage and answer the question 1.

What is technology doing to language? Many assume the answer is simple: ruining it. Kids can no longer write except in text-speak. Grammar is going to the dogs. The ability to compose thoughts longer than a tweet is waning.

Language experts tend to resist that gloom, noting that there is little proof that speech is really degenerating: kids may say "lol" out loud sometimes, but this is a marginal phenomenon. Nor is formal writing falling apart. Sentences like "omg wtf William teh Conqueror pwned Harold at Hastings in 1066!" tend to be written by middle-aged columnists trying to imitate children's supposed habits. A study by Cambridge Assessment, a British exam-setter, found almost no evidence for text-speak in students' writing.

Fortunately, the story of language and the internet has attracted more serious analysts, too. Now Gretchen McCulloch, a prolific language blogger and journalist — and herself of the generation that grew up with the internet — joins them with a new book, "Because Internet". Rather than obsessing about what the internet is doing to language, it largely focuses on what can be learned about language from the internet. Biologists grow bacteria in a Petri dish partly because of those organisms' short lifespans: they are born and reproduce so quickly that studies over many generations can be done in a reasonably short period. Studying language online is a bit like that: trends appear and disappear, platforms rise and fall, and these let linguists observe dynamics that would otherwise take too much time.

(248 words)

Question1: What's the author's attitude to the many people's assumption that Internet technology will ruin the language?

　　A) supportive.　　B) doubtful.　　C) disapproval.　　D) uncertain.

Passage Two

Read the following passage and answer the question 2.

  We have entered a new age of embedded, intuitive computing in which our homes, cars, stores, farms, and factories have the ability to think, sense, understand, and respond to our needs. It's not science fiction, but the dawn of a new era.

  Most people might not realize it yet, but we are already feeling the impact of what's known as the third wave of computing. In small but significant ways it is helping us live safer, healthier, and more secure lives. If you drive a 2014 Mercedes Benz, for example, an "intelligent" system endeavors to keep you from hitting a pedestrian. A farmer in Nigeria relies on weather sensors that communicate with his mobile device. Forgot your medication? A new pill bottle from AdhereTech reminds you via text or automated phone messages that it's time to take a pill.

  Technology is being integrated into our natural behaviors, with real-time data connecting our physical and digital worlds. With this dramatic shift in our relationship to technology, companies can adapt their products and services.

  We already see cities growing "smarter" by installing sensors to automate the management of parking spaces. To enhance urban security, acoustic sensors coupled with audio and GPS analytics "listen" to pinpoint the location of gunfire. Within 30 seconds, dispatchers can determine the number of shooters, the shots fired, and even the type of weapon used.

  Consider health care. Wearable devices allow us to monitor our steps, our sleep patterns, and our calorie intake to ensure we are following doctors' orders and meeting our personal goals. Parents of newborns can try a diaper that has a humidity sensor that tweets when it's time for a change.

  To understand how revolutionary the third wave is, we ought to consider how far we have come. The first wave began when companies started to manage their operations via mainframe computer systems over 50 years ago. Then computing got "personal" in the 1980s and 1990s with the introduction of the PC. For the most part, computing remained immobile and lacked contextual awareness.

(339 words)

计算机词汇:

intuitive computing 直观计算　　real-time data 实时数据

sensor *n.* (探测光、热、压力等的)传感器,敏感元件,探测设备

automated /ˈɔːtəmeɪtɪd/ *adj.* 自动化的; *v.* 使自动化 (automate 的过去式和过去分词)

GPS *abbr.* 全球定位系统 (Global Position System)

mainframe computer system 大型计算机系统,主机系统

Question 2: What's the author's attitude to the third wave of computing?

A) supportive.   B) doubtful.   C) disapproval.   D) uncertain.

Passage Three

Read the following passage and answer the question 3.

In 2006 Nick Bostrom of Oxford University observed that "once something becomes useful enough and common enough it's not labelled AI any more". Ali Ghodsi, boss of Databricks, a company that helps customers manage data for AI applications, sees an explosion of such "boring AI". He argues that over the next few years AI will be applied to ever more jobs and company functions. Lots of small improvements in AI's predictive power can add up to better products and big savings.

This is especially true in less flashy areas where firms are already using some kind of analytics, such as managing supply chains. When in September Hurricane Ian forced Walmart to shut a large distribution hub, halting the flow of goods to supermarkets in Florida, the retailer used a new AI-powered simulation of its supply chain to reroute deliveries from other hubs and predict how demand for goods would change after the storm. Thanks to AI this took hours rather than days, says Srini Venkatesan of Walmart's tech division.

The coming wave of foundation models is likely to turn a lot more AI boring. These algorithms hold two big promises for business. The first is that foundation models are capable of generating new content. Stability AI and Midjourney, two startups, build generative models which create new images for a given prompt. Request a dog on a unicycle in the style of Picasso — or, less frivolously, a logo for a new startup — and the algorithm conjures it up in a minute or so. Other startups build applications on top of other companies' foundation models. Jasper and Copy. AI both pay Open AI for access to GPT-3, which enables their applications to convert simple prompts into marketing copy.

(287 words)

计算机词汇：

analytics *n.* 分析法；解析法　　simulation *n.* [计] 仿真；模拟
foundation models　基础模型　　algorithm *n.*（尤指计算机）算法，运算法则
generative models　生成模型　　prompt *n.*（计算机屏幕上显示准备接受指令的）提示

Question 3: The author's attitude towards AI seems to be that of _____.

A) opposition.   B) suspicion.   C) approval.   D) indifference.

Unit Four

第四单元
阅读技巧：词义和句意推测

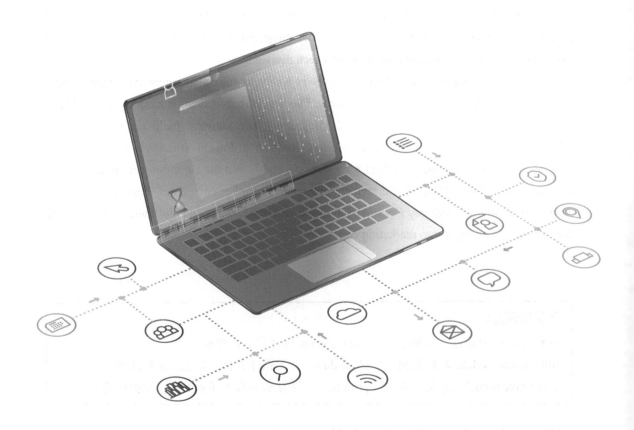

无论是在汉语还是英语阅读中，我们总会遇到生僻的字或难解的句子，从而影响我们对于文章的理解。如果这些词或句子不影响阅读理解，则可忽略；但如果他们对理解整个句子的意思或文章的主题至关重要，则必须借助猜测词义和推断句意的技巧攻克难关。

第一章 词义推测

第一节 策略讲解

猜测词义的技巧分为两类，即根据上下文和根据构词法知识猜测。

一、根据上下文猜词义

1. 根据定义或解释猜词义

在许多情况下，顾及读者的阅读体验，作者会在文章中对某个术语或生僻词给出该词的定义或用通俗易懂的词汇或句子进行解释。这种解释也可能是一种重复说明，近似于定义。通过阅读定义和解释部分，读者便可理解该词的基本意义。

以如下两句为例：

The great power of <u>tornadoes</u> is almost unbelievable. The speed of this whirling funnel-shaped wind may be more than 800 kilometers per hour.

即使读者不认识 tornado，也可从下文中"this whirling funnel-shaped wind"的解释推断出"漏斗状的高速旋转的风"是"旋风"或"龙卷风"。

Just like his <u>taciturn</u> father, Jon rarely says anything at family gatherings.

可以看出"taciturn"的词义和"rarely says anything"类似，所以前者的意思是沉默寡言，话少的。

2. 利用语法知识和标点符号猜词义

英语篇章中进行解释说明的语法现象颇多，如定语从句和同位语从句，甚至包括标点符号中冒号"："和破折号"——"表解释说明，分号"；"表并列，括号"( )"进行解释等的常见用法，都有助于读者对生词进行推测。

例如，George is a <u>scrooge</u>: he thinks only of money and will not spend a penny on anything he can get free. 解释型的标点符号——冒号后面的内容有助于帮助读者猜测"scrooge"的词义：他想到的只有钱，只要能够免费获得，就绝不多花一个子儿！显而易见，scrooge 作为与人搭配的表语，意思是"吝啬鬼，守财奴"。

再看同位语，Science has moved closer toward identifying the long-sought brain site of the "bodyclock", the timer that governs all the rhythms of life. 通过下文 body clock 的同位语"the timer …"及其定语从句解释了该短语的意思：人体内支配、影响生活节奏的计时器，可猜出 body clock 即生物钟。

练习猜测 ethics 的含义：Ethics — the standard of deeds and moral codes accepted by the society — has a powerful effect on modem business communication.

引出同位语的词语一般有 or、that is to say、in other words、namely 等。有的同位语是以括号或破折号的形式出现的，读者细心留意便很容易地找到同该词意义相近的词。

3. 根据上下文语境猜词义

通过上下文对相关信息进行联想，也可以帮助读者理解生词的含义。例如，He is a resolute man. Once he sets up a goal, he won't give it up easily. 从下文的细节阐述中，读者不难猜出，一个定下目标不轻易放弃的人就是"果断的""坚决的"人，因此 resolute 的含义一目了然。

4. 利用逻辑关系猜词义

所举的例子和例子要证明的观点相互构成解释，都有助于读者猜出生词的意思。例如，Some artists plan their paintings around geometric forms like squares, circles and triangles. 不认识 geometric 不要紧，因为根据 like 引出的几个例子(正方形、圆形、三角形)，概括出 geometric forms 的意思是"几何图形"。此类线索词包括 like、such as、for example、for instance、especially、including 等表明总分关系的词汇。

此外，在句子或段落中，如果两个事物或现象之间构成因果关系，便可根据因果关系推测生词词义。通过因果关系猜词，首先要找到生词与上下文之间表示因果关系的连接词，这样的连接词主要有：because、as、since、for、so、thus、as a result、of course、therefore 等。例如，Tom is considered an autocratic administrator because he makes decisions without seeking the opinions of others. 根据原因句"不听取别人的意见就做决定"进行推测，Tom 是个独断专行(autocratic)的人。

除了因果关系，比较或对比也常用来帮助词义猜测。文中用来比较的部分可以给读者提供理解不熟悉的词汇的线索。例如，The snow was falling. Big flakes drifted with the wind like feathers. 通过"like feathers"这样的比较，就可以猜出 flake 的意思：像羽毛一样随风飘落的雪自然就是"雪片""雪花"了。表示比较的引导词有 the same as...、like、similarly、likewise、in the same way/manner, equally 等。

当作者强调事物之间的区别和对立时，往往用对比的手法，表示对比的词语(往往是反义词)也可以暗示出生词的词义。例如，These shoes are stiff, but I like soft ones. 连词 but 表示转折和对比，熟词 soft (柔软的)与 stiff 形成对比，轻而易举就能猜出 stiff 是"僵硬的"。除 but 之外，表示对比关系的连接词还有 however、despite、not、but、unlike、in spite of、in contrast、whereas、while、on the contrary 等。

练习猜测划线单词的含义：

My son likes to take part in all kinds of social activities and make friends, while my daughter is introverted.

Diane experienced many great traumas during early childhood, including being injured in a car crash and the death of her family.

二、利用构词法知识猜词义

1. 辨认单词前/后缀和基本含义

英语单词的构词法有三种：派生、转化和合成。其中，派生词是通过在词根基础上，添加前缀（Prefix）和后缀（Suffix）构成新词。前缀是加在一个单词前面的音节，并不是独立的单词，不改变词性；但它本身具有一定的含义，能够改变原词的意思。因此，读者可以通过辨认前缀，并结合原词与前缀的意义猜出该词的新义。英文中表示反义的词缀很多，如 im-/in-/il-/ir-/un-/ab-/dis-/mis-等。uninteresting (un + interesting)前缀 un-表示"相反"，不难猜出这个词的意思是"不令人感兴趣的""无趣味的"。同样 maltreat (mal + treat)，如果记住前缀 mal-表示"恶""不良"，则可猜出该词的意思为"虐待"。

后缀一般也有一定的含义，能够改变单词的词性，有的也改变词义。根据认识的词根部分结合后缀的含义便能猜出生词的词义。例如 brokee (broke + ee)中-ee 是名词后缀，加在动词 broke (破产)之后，则表示"破产者"。同样 beautify (beauty + fy)中 beauty 意思是"美丽""美好的事物"，后缀-fy 同时具有构成动词的作用和表示"使……"的意义，因此，beautify 则表示"美化""打扮"。

转化法是指由一种词性转化为另一种或几种词性的方法。在这种构词法下，只要掌握某一词汇最常见的词性的含义，则不难猜测其它词性下的含义了。例如：

style 作为名词，意为 a general way of doing something 行为方式，风格，试猜测如下词性的含义：The dress is carefully styled.

可以断定，此处 style 为动词的被动语态形式，且与风格、款式相关，句子译为：这件连衣裙是精心设计的。可见，style 做动词，意为形成某种款式或样式(form a certain pattern)。

猜测如下划线词的词义：

The museum houses the biggest collection of antique toys in Europe.

She nursed her husband back to health.

He went upstairs to comfort her babies.

2. 分解复合词

复合词由两个或更多的词合成，故而可以从每一部分独立的意思猜出这个词的词义。如 skyscraper (sky + scraper)是一个复合词。第一部分 sky 表示天空；第二部分名词 scraper 是动词 scrape (擦，刮)的派生词。因此，这两个词的结合即比喻高得几乎刮到天的物体，即指"摩天大楼"。Lazybone (lazy + bone)是一个由 lazy (懒惰的)和 bone (骨头)构成的复合词，意思是"懒汉"或"懒虫"。

练习猜词义：

a. well-informed

b. peace-loving

c. man-made

综上，掌握必要的构词法知识，是学习英语词汇的必由之路。因此，应尽量多记常用的前缀及后缀，并掌握其意义，分析词汇构成的规律，以便扩大词汇量，从而提升阅读理解能力。

三、综合多种方法猜测词意

The criminal was to be killed at dawn; but he <u>petitioned</u> the king to save him and his request was granted. 通过上下文可知，被处死的人的请求(request)得到了国王的批准；那么可猜出 petitioned 就是"请求""恳求"的意思。这里是从同义词来猜生词词义的(在本句中 request 是名词，它的同源动词 request 与之同义)。

Doctors believe that smoking cigarettes is <u>detrimental</u> to your health. They also regard drinking as harmful. 从上下文看，detrimental 的意思同 harmful，在此，also 起到了提示作用；同时，从常识判断，吸烟没有好处，即有害处。当然，从构词上分析，名词 detriment (损害，有害)加后缀-al 变为形容词，即"有害的"。

Susan's dog is gentle and friendly; unfortunately, my dog doesn't have such a pleasant <u>temperament</u>. 从前一句看，gentle 和 friendly 是对性情、脾气的描写，可以推测 temperament 有此意，即"性情，脾性，气质"。从单词构成验证，名词 temper 意为"脾气""性情""气质"，加后缀-ment 仍为名词，意思相同。

练习猜测如下单词含义：

It is <u>absurd</u> to spend more money on highways. The wise solution for over crowded roads is public transportation.

The following Monday, when the chairman <u>convened</u> the second meeting of the committee, they also sat down quietly and waited for him to begin.

The government gave money to people to help buy homes outside of the cities. This system of <u>subsidized</u> housing caused many people to leave urban areas.

第二节 范文阅读

Passage One

Read the following excerpt and answer the question.

...It is pretty much a one-way street. While it may be common for university researchers to try their luck in the commercial world, there is very little traffic in the opposite direction.

Ex. By "a one-way street", the author means _____.

A) university researchers know little about the commercial world.

B) there is little exchange between industry and academia.

C) few industrial scientists would quit to work in a university.

D) few university professors are willing to do industrial research.

【答案与解析】C. 根据后文语境猜词义，下一句"对于大学研究人员来说，去商业领域碰运气的情况很常见，但反方向的情况很少"，后半句与前半句构成对比，言外之意是商业领域人士作为研究人员进入大学工作的情形很少，与 C 选项基本对等。

Passage Two

Read the following excerpt and answer the questions.

Internet fraud involves using online services and software with access to the internet to defraud or take advantage of victims. The term "internet fraud" generally covers cybercrime activity that takes place over the internet or on email, including crimes like identity theft, phishing, and other hacking activities designed to scam people out of money.

Internet scams that target victims through online services account for millions of dollars worth of fraudulent activity every year. And the figures continue to increase as internet usage expands and cyber-criminal techniques become more sophisticated.

In the USA Internet fraud offenses are prosecuted under state and federal law. For example, federal law has the controlling statute U.S.C.1343 that covers general cyber fraud and can carry a punishment of up to 30 years in prison and fines of up to $1 million depending on the severity of the crime.

States like California also have anti-phishing, credit card fraud, unauthorized computer access, and identity theft laws. These laws also prohibit eliciting personally identifiable information (PII) via the internet by pretending to be a company under the Anti-Phishing Act of 2005.

(182 words)

计算机词汇：

internet fraud　网络诈骗　　　　cybercrime activity 网络犯罪活动
identity theft 身份盗用

Ex.1. What is the possible meaning of "phishing"?

　　A) Fishing for food.　　　　　　B) Crime of fishing private information for benefits.
　　C) Prohibiting.　　　　　　　　D) Robbery.

【答案与解析】B. 利用构词法及单词发音（同 fish），再结合前文 including crimes like 由一般到特殊，所以可以断定 phish 指的是一种网络犯罪，答案为 B 或者 D，再联想常识 fish 的常用义，此处 phish 意思是网络钓鱼。

Ex.2. The word "prosecuted" could best be replaced by_____.
A) proved.　　　　B) controlled.　　　　C) supported.　　　　D) charged.

【答案与解析】D. 由动词的主语 Internet fraud offense 和动词发出者 law 可以判断，prosecuted 与法律相关，下文中 punishment 和 fines（罚款）进一步验证了答案。

Passage Three

Read the following excerpt and answer the question.

　　In spite of "endless talk of difference", American society is an amazing machine for homogenizing people. This is "the democratizing uniformity of dress and discourse, and the casualness and absence of consumption" launched by the 19th – century department stores that offered vast arrays of goods in an elegant atmosphere. Instead of intimate shops catering to a knowledgeable elite, "these were stores "anyone could enter, regardless of class or background. This turned shopping into a public and democratic act. The mass media, advertising and sports are other forces for homogenization.

(89 words)

Ex. The word "homogenizing" most probably means_____.
A) identifying.　　　B) association.　　　C) assimilating.　　　D) monopolizing.

【答案与解析】C. 前缀 homo-表示"同一"，例如 homosexual = homo(同一) + sexual(性的)：同性恋的，可以初步判断该词的含义与"相同"有关。另外，后文主要在讨论美国的社会机制使得来自不同国家和地区的人都变得同一。因此我们可以得出，"homogenizing"的含义为"同化"，选项中与其同义替换的词为"assimilating"。

第三节　阅读演练

Passage One

Read the following excerpt and answer the questions 1-2.

　　Private firms should in general be free to deal with whom they like. Just as Facebook may legally ban people like Donald Trump from its network and Amazon Web Services can decline to host alt-right-friendly apps like Parler, Mastercard is free to drop particular unhealthy sites.

Yet the market power of these firms and their often like-minded rivals means that without their approval and support, individuals or businesses may face exclusion, even if they have broken no law.

(78 words)

计算机词汇：

Facebook 脸书   Amazon 亚马逊公司   social network 社交网络

Question 1: What's the possible meaning of "alt-right-friendly apps"?

    A) Apps which are always right.

    B) Apps which are friendly to users.

    C) Apps which are favorable to people believing in absolute Right belief.

    D) Parler declined by Amazon Web Services.

Question 2: Which word can be used to replace "approval" (Line4)?

    A) Rival.   B) Opposition.   C) Proof.   D) Favorable opinion.

Passage Two

Read the following passage and answer the question 3.

    The pandemic embedded video into the workplace. Workers who had never previously been on camera suddenly spent every hour of the day getting used to the sight of themselves and their colleagues on screen. Executives realized that they could send video messages to their workforces rather than having to convene town halls.

    There is no going back. Blogs have become vlogs. Meetings are now recorded as a matter of course, so that people can fail to watch them back later. Some firms routinely ask applicants to record answers to certain questions on video, so that people can see how well prospective recruits communicate. Since video has become more central to work, it pays to be good at it. Being a star in the video age means having the right set-up, speaking well and listening well.

(135 words)

Question 3: What is the possible meaning of "prospective"?

Passage Three

Read the following passage and answer the questions 4-6.

China's internet overseer laid out regulations recently to prevent <u>minors</u> from spending too much time on their smartphones, dealing a potential blow to Tencent and other social media leaders.

The regulations published by the Cyberspace Administration of China represent some of the harshest restrictions on internet use in the world, as worries fester about online addiction. "Non-adult" children won't be allowed to access the internet from mobile devices from 10 p.m. to 6 a.m., the agency said in a draft of rules published on its website. Other restrictions include a maximum of two hours' mobile usage for those 16 to 18.

Beijing since 2021 has pursued campaigns to alleviate the burden on minors and their families and combat what it views as social ills. It imposed limits on online gaming for kids and declared for-profit after-school tutoring illegal, measures regarded as intended to both lessen financial burdens and promote healthier activities. At one point in 2021, state-backed media referred to gaming as "spiritual opium."

Tencent's WeChat and ByteDance's TikTok-like Douyin are among the most popular and heavily used internet services in China, attracting a disproportionate number of minors. The CAC didn't name any services except to say that platforms will be responsible for ensuring they meet the CAC's requirements, which include promoting lullabies for children under 3 and educational news and entertainment content to those under 12.

Chinese technology shares were down in the late afternoon in Hong Kong, mostly extending losses from the morning. Kuaishou Technology, a short video service, slid about 4%, while social media app Weibo Corp. <u>shed</u> more than 5% of its value. Tencent fell more than 3%. ByteDance, the world's most valuable startup, isn't publicly traded.

The regulator said platform providers will be responsible for enforcement, although it didn't specify penalties for violations. Companies in China are typically held accountable for implementing such government regulations. Under previous rules limiting game play for minors, companies required real-name registrations for users and then introduced technology to cut off customers <u>outside of authorized times</u>.

(336 words)

计算机词汇：

Tencent 腾讯　　　　　　WeChat 微信　　　　　　ByteDance 字节跳动科技有限公司
Cyberspace Administration of China（CAC）：国家互联网信息办公室
online addiction 网络上瘾　　　　　online gaming 网络游戏

| for-profit after-school tutoring 课后营利性补习 | TikTok-like Douyin (国内)抖音 |
| Kuaishou Technology 快手科技有限公司 | real-name registration 实名认证 |

Question 4: The word "minors" most probably means_____.
    A) Kids.   B) Subjects.   C) Children under 12.   D) "Non-adult" Children.

Question 5: What could be the possible meaning of "shed" (Para.5)?
    A) Increased. B) Dropped.   C) Reached.   D) Limited.

Question 6: What does "outside of authorized times" refer to?
    A) Recognized by authority.   B) Beyond Specified times.
    C) Not in the author's age.   D) Without limited by rules.

第二章　句意推测

第一节　策略讲解

有时，为了某种目的，作者往往不直接说出某一意思，而是含蓄地表达。这种隐含的意思有时是篇章的中心思想，有时是个别重点句子，所以阅读短文经常需要对句子进行推论 (making inference)。所以，从范围大小看，推理分为全文和局部推理；从类型看，推理一般包括数字推理、知识推断和逻辑推理。

一、全文推理：针对全文段落主旨和作者的观点态度的推理。读者必须学会找到段落主题或者论点的句群，并归纳原文中作者的态度。

读文章，回答问题：

In recent years, Israeli consumers have grown more demanding as they've become wealthier and more worldly-wise. Foreign travel is a national passion this summer alone, one in 10 citizens will go abroad. Exposed to higher standards of service elsewhere, Israelis are returning home expecting the same. American firms have also begun arriving in large numbers. Chains such as KFC, McDonald's and Pizza Hut are setting a new standard of customer service, using strict employee training and constant monitoring to ensure the friendliness of frontline staff. Even the American habit of telling departing customers to "Have a nice day" has caught on all over Israel. Nobody wakes up in the morning and says, 'Let's be nicer," says Itsik Cohen, director of a consulting firm. "Nothing happens without competition."

　　Privatization for the threat of it, is a motivation as well. Monopolies (垄断者)that until recently have been free to take their customers for granted now fear what Michael Perry, a marketing professor, calls "the revengeful (报仇的) consumer". When the government opened up competition with Bezaq, the phone company of its international branch lost 40% of its market share, even while offering competitive rates. Says Perry, "People wanted revenge for all the years of bad service?" The electric company whose monopoly may be short-lived has suddenly stopped requiring users to wait half a day for a repairman. Now, appointments are scheduled to the half-hour. The graceless £1A1 Airlines, which is already at auction (拍卖)，has retrained its employees to emphasize service and is boasting about the results in an ad campaign with the slogan "You can feel the change in the air." For the first time, praise out numbers complaints on customer survey sheets.

Question: It may be inferred from the passage that_____.

　　A) customer service in Israel is now improving.

　　B) wealthy Israeli customers are hard to please.

　　C) the tourist industry has brought chain stores to Israel.

　　D) Israeli customers prefer foreign products to domestic ones.

【答案与解析】由短文主题 customer service 及第一段要点：以色列服务提升的原因和做法举例，不难推出答案为 A。

　　可见，对全文进行推理需要读者总结文章主题和作者态度，结合字里行间的细节。

二、局部推理：分为根据给定的段落进行推理和根据某个信息点进行推理。

以下例子为某个信息点推理：

　　Fortunately, the White House is starting to pay attention. But it's obvious that a majority of the president's advisers still don't take global warming seriously. Instead of a plan of action, they continue to press for more research — a classic case of "paralysis by analysis".

Question: What does the author mean by "paralysis by analysis"?

　　A) Endless studies kill action.

　　B) Careful investigation reveals truth.

　　C) Prudent planning hinders progress.

　　D) Extensive research helps decision-making.

【答案与解析】这类推理首先需要返回原文，找到该句，通过它的前一句和后一句揭示谜底，读者只需根据上下文逻辑关系和语义推理该句句意。本句中，instead of a plan of action 的对比用法，揭示了白宫的总统顾问们只讨论，不采取行动的做法，endless studies 与 more research 含义相同，所以答案是 A。

　　In the weeks and months that followed Mr. Hirst's sale, spending of any sort became deeply unfashionable. In the art world that meant collectors stayed away from galleries and salerooms.

Sales of contemporary art fell by two-thirds, and in the most overheated sector, they were down by nearly 90% in the year to November 2008. Within weeks the world's two biggest auction houses, Sotheby's and Christie's, had to pay out nearly $200m in guarantees to clients who had placed works for sale with them.

Question: By saying "spending of any sort became deeply unfashionable", the author suggests that _____.

 A) collectors were no longer actively involved in art-market auctions.

 B) people stopped every kind of spending and stayed away from galleries.

 C) art collection as a fashion had lost its appeal to a great extent.

 D) works of art in general had gone out of fashion so they were not worth buying.

【答案与解析】首先定位回原文之后，发现定位句中既没有指代表示与上文的联系，也没有其他语义标志。但定位句的位置在段落首句，因此我们可以利用总分的语义一致关系，利用下文理解定位句。下一句中出现了指代"that"表示与上文的联系，其次还出现了重要的"meant"，由此我们可以确定下一句和我们题干定位句的语义一致关系，只需要理解下一句就可解题。"In the art world that meant collectors stayed away from galleries and salerooms."翻译为"在艺术界，这意味着收藏家远离画廊和拍卖场。"这与选项 A 匹配，collector 是原词，no longer actively involved≈stay away，art-market auctions 约等于 galleries and salerooms，是和原文的同意替换加原词复现。

 此外，读者的阅历和作者的语气、语调、措辞等文体特征，也可帮助读出作者的"言外之意"。总之，要在原文基础上进行有理有据的推断，断章取义或者推理过度都会导致错误的结论。

第二节 范文阅读

Passage One

Read the following passage and answer the question.

 Some of the concern over the violence of computer games has been about children who are unable to tell the difference between fiction and reality and who act out the violent moves of the games in fight on the playground. The problem with video games is that they involve children more than television or films and this means there are more implications for their social behavior. Playing these games can lead to anti-social behavior, make children aggressive and affect their emotional stability.

(82 words)

Ex. According to the author, why do video games lead to violence more than TV or movies?

_____.

【答案与解析】通过关键词 video games、television or films，精准定位关键句 The problem with video games is that they involve children more than television or films and this means there are more implications for their social behavior. 电子游戏比电影或电视都更容易让孩子们有身临其境的参与感，从而对他们的社会行为影响更大。题目中"暴力"便是社会行为的一种，总分逻辑关系说明答案为：Because video games involve children more.

Passage Two

Read the following passage and answer the question.

  Imagine you went to a restaurant with a date; had a burger, paid with a credit card, and left. The next time you go there, the waiter or waitress, armed with your profile data, greets you with, "Hey Joe, how are you? Mary is over there in the seat you sat in last time. Would you like to join her for dinner again?" Then you find out that your burger has been cooked and your drink is on the table. Forget the fact that you are with another date and are on a diet that doesn't include burgers. Sound a little bizarre? To some, this is restaurant equivalent of the Internet. The Net's ability to profile you through your visits to and interactions at websites provides marketers with an enormous amount of data on you — some of which you may not want them to have.

  Are you aware that almost every time you access a website you get a "cookie"? A cookie on the Internet is a computer code sent by the site to your computer — usually without your knowledge. During the entire period of time that you are at the site, the cookie is collecting information about your interaction...

(197 words)

计算机词汇：

profile 从侧面介绍   marketer 生意人；商人
cookie 网络跟踪器（记录上网用户信息的软件） computer code 计算机代码

Ex. What's Joe's possible attitude towards the waiter or waitress?

_____.

【答案与解析】第 1 段中，乔另约了一名新女友去餐馆，餐馆服务员却招呼他坐在上次约会的女友身边，并且为他准备了与上次同样的食品，包括他节食忌食的汉堡，这样的服务肯定会使乔做出负面的评价。此外，后文中，bizarre, some of which you may not want them to have 泄露了作者对滥用别人的个人资料的反感，因此可推断乔对餐厅服务员的做法会感到生气。

Passage Three

Read the following passage and answer the question.

  Instead, the key to making payments more competitive in America is to create a new network of financial plumbing: a "real-time" interbank-payment system allowing for near-instant and cheap transfers. Swathes of Europe and Asia have already done this. Once this exists, banks and fintechs can build products, standards and services on top of it. In Singapore and the Netherlands, for example, those efficient payment pipes are open to digital wallets, which can process payments in a few clicks, taps or by scanning a QR code.

(67 words)

计算机词汇：
"real-time" interbank-payment 实时跨行支付   fintech 金融科技公司
digital wallet 数字钱包   QR code 二维码

Ex. How can the financial plumbing benefit payments in America?

_____.

【答案与解析】本推理题需要进行归纳。第一句中阐述了金融管道网络使美国的支付方式更具有竞争力，后面的冒号揭示了前后句之间解释说明的关系，即金融管道的益处在于 a "real-time" interbank-payment system allowing for near-instant and cheap transfers。此外，后文中新加坡和荷兰的例子也说明金融管道网络的另一益处，即简单、易操作。

Passage Four

Read the following passage and answer the question.

  Before ChatGPT came along, an economics teacher might ask pupils to write an essay describing Keynesianism.With ChatGPT as an option, the teacher might ask the students to assess and revise the chatbot's response to the same question — a more difficult task. AIs have other practical uses for teachers. They can help write lesson plans and worksheets at different reading levels and even in different languages. They can also cut down the time spent on duties, such as writing recommendation letters, that devour time that could be spent teaching.

Some organizations are going even further. Khan Academy, an education non-profit, recently launched a pilot of Khanmigo, its virtual guide that uses GPT-4, the latest upgrade of ChatGPT, to support pupils and teachers.

(117 words)

计算机词汇：

ChatGPT　OpenAI 研发的一款聊天机器人程序　　　　　chatbot　聊天机器人
Khan Academy　可汗学院　　virtual guide　虚拟向导/导游　　upgrade　提升；升级版

Ex. Despite its demerits, ChatGPT is useful for teaching. Why does the author list the economics teacher's assignment as an example?

_____.

【答案与解析】从本段第三行 AIs have other practical uses for teachers 中的 other 说明前几句阐述了 AI 的至少一个用处。所以，第一个例子是为了举例说明 ChatGPT 对于教学的帮助，而例子并非平铺直叙，而是通过使用 ChatGPT 前、后教师布置作业的对比来证明它的用处。

Passage Five

Read the following passage and answer the questions.

Over the past ten months most people in the rich world have participated in the biggest shopping revolution in the West since malls and supermarkets conquere suburbia 50 years ago. The pandemic has led to a surge in online spending, speeding up the shift from physical stores by half a decade or so. Forget the chimney; Christmas gifts in 2020 came flying through the letterbox or were dumped on the doorstep. Workers at a handful of firms, including Amazon and Walmart, have made superhuman efforts to fulfil online orders, and their investors have made supernormal profits as Wall Street has bid up their shares on euphoria that Western retailing is at the cutting edge.

Yet as we explain this week it is in China, not the West, where the future of e-commerce is being staked out. Its market is far bigger and more creative, with tech firms blending e-commerce, social media and razzmatazz to become online-shopping emporia for 850m digital consumers. And China is also at the frontier of regulation, with the news on December 24th that trustbusters were investigating Alibaba, co-founded by Jack Ma, China's most celebrated tycoon, and until a few weeks ago its most valuable listed firm . For a century the world's consumer businesses have looked to America to spot new trends, from scannable barcodes on Wrigley's gum in the 1970s to keeping up with the Kardashians' consumption habits in the 2010s. Now they should be looking to the East.

China's lead in e-commerce is not entirely new. By size, its market overtook America's in 2013 — with little physical store space, its consumers and retailers leapfrogged ahead to the digital world. When Alibaba listed in 2014 it was the world's largest-ever initial public offering. Today the country's e-retailing market is worth $2trn, more than America's and Europe's combined. But beyond its sheer size it now stands out from the past, and from the industry in the West, in several crucial ways.

(324 words)

计算机词汇：

online spending 网上消费　　　　　　physical stores 实体店

fulfil online orders 完成在线订单　　　e-commerce 电子商务

scannable barcode 可扫描条形码　　　initial public offering 首次公开发售

online-shopping emporia 在线购物商城

Ex.1. Which will not be the possible result of the biggest shopping revolution in the West?

　　A) Malls and supermarkets can be found everywhere in and out of cities.

　　B) Online spending is increasing dramatically.

　　C) Physical stores are at low tide in advance.

　　D) Christmas gifts will experience door-to-door delivery.

Ex.2. What advantages does Chinese e-commerce particularly hold while western countries don't harbor?

　　A) Workers have to make efforts to fulfil online orders.

　　B) Its market is far bigger and more creative.

　　C) Their investors have made an enormous profits in the past years.

　　D) Tech firms in China have leapfrogged ahead in China.

Ex.3. Why does the author believe American consumer businesses have to turn their attention to the east in the future?

　　A) Because American consumption habits have been changed.

　　B) Because China has taken the place of the West as the largest market for its unique futures.

C) Because the West has been under the influence of pandemic.

D) Because Jack Ma co-founded Alibaba.

【答案与解析】

Ex.1. A. 本题考查推理技巧的掌握：购物剧变可能带来的后果。第一句中"50 年前商场和超市开始在郊区大受欢迎"说明实体店受欢迎是 50 年前，而不是现在购物剧变后，所以 A 答案不是结果，符合问题要求。而 B,C,D 选项均是线上购物带来的变化，不符合题干要求。

Ex.2. B. 本题考查细节：中国电子商务具备的、西方没有的特点。第二段第一句揭示了二者之间的对比关系。A 选项处理订单的情形中外电子商务领域都有，C 选项盈利大在第一段中明言属于美国电子商务，常理推断属于中西共有特点，D 选项切忌因 tech firms 而断章取义，尽管原文中中国的科技公司有被提及，但原文指的是科技公司与电子商务相结合构成中国的特点之一，而不是科技公司本身的迅猛发展。采用排除法，答案为 B。

Ex.3. B. 本题需要在把握细节的基础上对全文进行适当推理。首先，破解题干。问题关于美国消费领域将注意力转向东方的原因，根据全文逻辑发展，从第二段开始东方世界电子商务领域发展的代表就是中国；其次，归纳原文中中国电子商务领域的发展成就，答案 A 最具有概括性。

第三节　阅读演练

Passage One

Read the following passage and answer the question 1.

When Steve Jobs unveiled the iPhone in 2007, he changed an industry. Apple's brilliant new device was a huge advance on the mobile phones that had gone before: it looked different and it worked better. The iPhone represented innovation at its finest, making it the top-selling smartphone soon after it came out and helping to turn Apple into the world's most valuable company, with a market capitalization that now exceeds $630 billion.

Apple's achievement spawned a raft of imitators. Many smartphone manufacturers now boast touch-screens and colourful icons. Among them is Samsung, the world's biggest technology manufacturer, whose gadgets are the iPhone's nearest rivals and closest lookalikes. The competition and the similarities were close enough for Apple to sue Samsung for patent infringement in several countries, spurring the South Korean firm to counterclaim that it had been ripped off by Apple as well. On August 24th an American jury found that Samsung had infringed six patents and ordered it to pay Apple more than $1 billion in damages, one of the steepest awards yet seen in a patent case.

Some see thinly disguised protectionism in this decision. That does the jury a disservice: its members seem to have stuck to the job of working out whether patent infringements had occurred. The much bigger questions raised by this case are whether all Apple's innovations should have been granted a patent in the first place; and the degree to which technology stalwarts and start-ups alike should be able to base their designs on the breakthroughs of others.

It is useful to recall why patents exist. The system was established as a trade-off that provides a public benefit: the state agrees to grant a limited monopoly to an inventor in return for disclosing how the technology works. To qualify, an innovation must be novel, useful and non-obvious, which earns the inventor 20 years of exclusivity. "Design patents", which cover appearances and are granted after a simpler review process, are valid for 14 years.

(330 words)

计算机词汇：

market capitalisation 市场总值　　touch-screen 触摸屏　　icon 图标
Samsung 三星集团　　　　　　　　patent infringement 专利侵权

Question 1: Which is not true about the patents of mobile phone?
　　A) The touch-screens and colourful icons of Samsung are similar to those of Apple, which led to their fight in patents.
　　B) Apple obviously should keep the patent for all its creativity.
　　C) Design patents in appearance can win less few years for its inventor than those in utility.
　　D) Technological firms are not certain to which extent they can make use of the design innovation of others.

Passage Two

Read the following passage and answer the question 2.

The number of manufacturers at the chip making industry's cutting-edge has fallen from over 25 in 2000 to three. The most famous of that trio, Intel, is in trouble. It has fired its boss, an admission that it has fallen behind. It may retreat from making the most advanced chips, known as the three-nanometre generation, and outsource more production, like almost everyone else. That would leave two firms with the stomach for it: Samsung in South Korea and TSMC in

Taiwan. TSMC has just announced one of the largest investment budgets of any private firm on the planet. An array of corporate A-listers from Apple and Amazon to Toyota and Tesla rely on this duo of chipmakers.

The other big industry rupture is taking place in China. As America has lost ground in making chips, it has sought to ensure that China lags behind, too. The American tech embargo began as a narrow effort against Huawei over national security, but bans and restrictions now affect at least 60 firms, including many involved in chips. SMIC, China's chip champion, has just been put on a blacklist, as has Xiaomi, a smartphone firm. The cumulative effect of these measures is starting to bite. In the last quarter of 2020 TSMC's sales to Chinese clients dropped by 72%.

In response, China is shifting its macro-control button into its highest gear in order to become self-sufficient in chips faster. Although chips have featured in government plans since the 1950s it is still five to ten years behind. A $100bn-plus subsidy kitty is being spent freely: last year over 50,000 firms registered that their business was related to chips — and thus eligible. Top universities have beefed up their chip programmes. If the era of advanced chips being made in America may be drawing to a close, the age of their manufacture in China could be beginning.

(310 words)

计算机词汇：

TSMC 台积公司　　　　Toyota 丰田汽车　　Tesla 特斯拉汽车
embargo 禁令　　　　　SMIC 中芯国际

Question 2: We can infer about chip making industry except_____.

A) samsung and TSMC will gain some market shares in the production of three-nanometre generation from Intel.

B) some of Chinese firms involving chips like SMIC and Xiaomi have been suppressed by America.

C) those bans and restrictions from America against Chinese firms have no effect on the former.

D) chinese governments, companies as well as universities are exerting themselves for the realization of self-reliance in chips.

Passage Three

Read the following passage and answer the question 3.

How much have you spent on the cloud today? It takes Robert Hodges only a few clicks to find out. He pulls up a dashboard on a computer in his home office in Berkeley, California, which shows cloud spending at his database firm, Altinity, in real time. The cloud represents half of Altinity's total costs.

Mr Hodges's widget is a window onto the future. As bills soar, every firm of any size will need to understand not just the benefits of the cloud, but also its costs. Gartner, a consultancy, calculates that spending on public-cloud services will reach nearly 10% of all corporate spending on information technology (IT) in 2021, up from around 4% in 2017. Plenty of technophile startups spend 80% of their revenues on cloud services, estimate Sarah Wang and Martin Casado of Andreessen Horowitz, a venture-capital firm. The situation is analogous to a century ago, when electric power became an essential input (and prompted some firms to hire another kind of CEO: the chief electricity officer).

For cloud companies this has been a bonanza. Giants of the industry, such as Amazon Web Services (AWS), Microsoft Azure, Google Cloud Platform (GCP) and, in China, Alibaba and Tencent, have been adding business briskly. Gartner expects global sales of cloud services to rise by 26% in 2021, to more than $400bn. But competition is stirring. On December 9th Oracle, a big software-maker, reported higher revenue than expected, mainly thanks to the rapid growth of its cloud unit. Its market value shot up by over 15%, or nearly $40bn. And a welkin of companies is emerging to help businesses manage their computing loads. One such firm, Snowflake, is worth $108bn. Another, HashiCorp, went public in New York on December 8th and now boasts a stockmarket value of $15bn, three times its last private valuation in 2020.

(305 words)

计算机词汇：

dashboard 仪表盘　database 数据库　　technophile 技术爱好者
Oracle 甲骨文公司　market value 市值　public-cloud service 公共云服务

Question 3: What does the author think of cloud services?

　　A) It's challenging.　　　　　　　B) It's promising.
　　C) It's adventuring.　　　　　　　D) Its future is uncertain.

Unit Five
Network Techniques

— 第五单元 —

网络技术

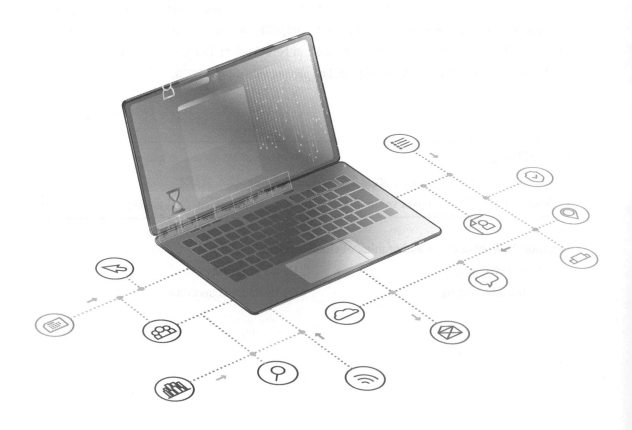

第一章　主题阅读

Text

The Internet's Next Act

1　In 2007 more humans lived in cities than outside them for the first time. It was a transition 5,000 years in the making. The internet has been quicker to reach the halfway mark. Over 50% of the planet's population is now online, a mere quarter of a century after the web first took off among tech-savvy types in the West. The second half of the internet revolution has begun. It is changing how society works — and also creating a new business puzzle.

2　Most new users are in the emerging world; some 726m people came online in the past three years alone. China is still growing fast. But much of the rise is coming from poorer places, notably India and Africa. Having seen what fake news and trolling has done to public discourse in rich countries, many observers worry about politics being debased, from the polarization of India's electorate to the persecution of Myanmar's Rohingya minority. On the positive side, charities and aid workers talk endlessly and earnestly about how smartphones will allow farmers to check crop prices, let villagers sign up for online education and help doctors boost vaccination rates.

3　Less well appreciated is that the main attractions of being online are the same for the second half as they were for the first. Socializing and play, not work and self-improvement, are the draw. Messaging apps help friends stay in touch, and let migrant workers say goodnight to their children back home. People entertain their friends — and strangers — on social media with goofy home-made videos on YouTube or TikTok, an app focused on short, humorous clips. Cheap data plans and thumb drives bring pirated films to millions who may never have been to a cinema. Dating apps are more popular than farming advice; video games are more popular than either. Such boons are unlikely to make their way into many UN development reports. But they are a boost to the stock of human happiness.

4　For businesses, the second half of the internet offers a vast pool of customers. It also brings a headache — most of these new users are not rich enough. Tens of billions of dollars in venture-capital money have flowed into internet startups in emerging markets. The Silicon Valley giants have built up big user bases — over 1.5bn Facebook users are in developing countries. YouTube, a video site owned by Google, is increasingly dominated by non-Western users. Last year Walmart spent $16bn buying Flipkart, an Indian e-commerce giant. Jumia, an e-commerce firm with 4m customers in Nigeria and 13 other African countries, floated in New York in April.

5　Despite these firms' punchy valuations, they are still looking for sustainable business models. Reliance Jio, an Indian firm, has sunk $37bn into building a high-speed mobile network and acquiring a big base of mostly poor users. Each Facebook user in Asia generates only $11 of advertising revenue a year, compared with $112 for a North American one. The combined revenue of all the internet firms in emerging markets (excluding China) is perhaps $100bn a year. That is about the same size as Comcast, America's 31st-biggest listed firm by sales.

6　Nonetheless, the impact of these firms on business will get bigger in two ways. First, they will grow fast — although whether fast enough to justify their valuations remains to be seen. To maximise their chances, many are offering not just a single service (such as search or video), as Western firms tended to in their early years, but a bundle of services in one app instead, in the hope of making more money per user. This approach was pioneered in China by Alibaba and Tencent. Go-Jek in Indonesia offers ride-hailing, payments, drug prescriptions and massages. Facebook is pushing a digital payments system in India through its chat service, WhatsApp.

7　The second is that in the emerging world, established firms are likely to be disrupted more quickly than incumbents were in the rich world. They have less infrastructure, such as warehouses and retail sites, to act as a barrier to entry. Many people, especially outside the big cities, lack access to their services entirely. Beer, shampoo and other consumer-goods firms could find that as marketing goes digital, new insurgent brands gain traction faster. Banks will be forced to adapt quickly to digital payments or die. Viewed this way, there is a huge amount of money at stake — the total market value of incumbent firms in the emerging world, outside China, is $8trn. If you thought the first half of the internet revolution was disruptive, just wait until you see the second act.

The Economist 2019-06-09

Vocabulary

acquire *v.* 获得，得到；购得
at stake *adv.* 危如累卵，危险
boon *n.* 非常有用的东西；益处
debase *v.* 降低质量；贬值
disrupt *v.* 扰乱；使中断
emerge *v.* 出现，兴起；摆脱；暴露
earnestly *adv.* 认真地，诚挚地

generate *v.* 产生，引起
impact *n.* 影响；作用；冲击力；*v.* 对⋯有影响；冲击；撞击
infrastructure *n.* 基础设施(建设)
incumbent *adj.* 在职的；义不容辞的；*n.* 在职者；教会中的任职者
insurgent *n.* 起义者，叛乱者

maximise *v.* 最大限度地利用
justify *vt.* 证明…有理；为…辩护；对…作出解释
migrant *n.* 候鸟；移居者，移民；随季节迁移的工作者；迁移动物
pirate *n.* 海盗，强盗；海盗船；侵害版权者；盗版者；*vt.* 抢劫；剽窃，盗用
planet *n.* 行星；地球
prescription *n.* [医]药方，处方；指示；法规

punchy *adj.* 简洁有力的；言简意赅的；生气勃勃的；有分量的
revenue *n.* 税收；财政收入；收益
startup *n.* 启动；新兴公司，新开张的企业
sustainable *adj.* 可持续的；可以忍受的；可支撑的
transition *n.* 过渡，转变，变迁
traction *n.* 拖拉；牵引力；摩擦力
vaccination *n.* [医]种痘，接种
venture-capital 风险投资

Cultural Notes

Myanmar Officially the Republic of the Union of Myanmar, also known as Burma, is a country in Southeast Asia. It is the largest country by area in Mainland Southeast Asia, and has a population of about 54 million as of 2017. It is bordered by Bangladesh and India to its northwest, China to its northeast, Laos and Thailand to its east and southeast, and the Andaman Sea and the Bay of Bengal to its south and southwest. The country's capital city is Naypyidaw, and its largest city is Yangon.

YouTube Social media platform and website for sharing videos, headquartered in California. It was officially launched on December 15, 2005, by Steve Chen, Chad Hurley, and Jawed Karim, three former employees of the American e-commerce company PayPal. It was serving more than two million video views each day since then.

TikTok A popular social media app that allows users to create, watch, and share 15-second videos shot on mobile devices or webcams. With its personalized feeds of quirky short videos set to music and sound effects, the app is notable for its addictive quality and high levels of engagement.

Silicon Valley A region in the south San Francisco Bay Area. The name was first adopted in the early 1970s because of the region's association with the silicon transistor, which is used in all modern microprocessors. The area is notable as a global hub for the vast number of technology companies that are headquartered there. It is also known for being a center for innovation, entrepreneurial spirit, and a lifestyle founded on technologically based wealth.

Walmart Also known as: Wal-Mart, Wal-Mart Stores, Inc., Walmart, Inc., American operator of discount stores that was one of the world's biggest retailers and among the world's largest corporations. Company headquarters are in Bentonville, Arkansas, founded by Sam Walton in Rogers, Arkansas, in 1962. An emphasis on customer attention, cost controls, and efficiencies in its distribution networks (e.g., regional warehousing) helped Wal-Mart become the largest retailer in the United States in 1990. It moved into international markets one year later with the opening of a store in Mexico, and later in Canada, China, Germany, and the United Kingdom.

WhatsApp At its most basic level, it's simply a chat app for exchanging messages with friends, not unlike the SMS text messaging that's built into nearly every mobile phone. It was acquired by Facebook (now Meta) in 2014, with over 5 billion installs from the Google Play Store and 2 billion active monthly users.

Read between Lines

Direction: Read the passage carefully and answer the questions below.

Question 1: How has the Internet developed since it took off in the past twenty years? (Para.1)

_____.

Question 2: Where do quite a few new Internet users come from globally? (Para.2)

_____.

Question 3: What positive effects has Internet had on its users according to the passage? (Para.2)

_____.

Question 4: What socializing apps have been mentioned to prove the second half's attraction? (Para.3)

_____.

Question 5: What have those non-western Internet companies done in order to increase their values? (Para.6)

_____.

Question 6: Why do those online companies are more easily faced with failure compared with present ones? (Para.7)

_____.

Words and Expressions

Ex.5.1. *Write the word according to the definition. The first letter is given.*

1. e_____ : come out into view.
2. f_____ : something that is a counterfeit; not what it seems to be.
3. m_____ : traveler who moves from one region or country to another.
4. r_____ : the entire amount of income before any deductions are made.
5. j_____ : show to be reasonable or provide adequate ground for.
6. t_____ : a change from one place or state or subject or stage to another.

Ex.5.2. *Fill in the blanks of the appropriate word from the box below. Change the form if necessary.*

| impact | generate | pioneer | appreciate | earnestly |
| sustainable | dominate | entertain | transition | giant |

1. We _____ hope what we learned will serve us well in our new job.
2. It is expected that the meeting will have a marked _____ on the future of the country.
3. I have some confidence that houses will _____ in value in the next five years.
4. A _____ way is the cornerstone (基石) of our nation's energy security, and will be one of the major challenges of the 21st century.
5. The computer bug is _____ chaos in the office.
6. Stage shows were laid on to _____ the foreign guests.
7. Renewables account for only a small share of global primary energy consumption, which is still _____ by fossil fuels — 30% each for coal and oil, 25% for natural gas.
8. He will remain head of state during the period of _____ to democracy.
9. Professor Alec Jeffreys invented and _____ DNA tests.
10. An investigation found that media and telecom _____ Comcast is the most hated provider.

Ex.5.3. *Complete each sentence below with the appropriate form of word in bracket.*

1. Thousands were forced to _____ (migrant) from rural to urban areas in search of work.

2. It is to recognize that in America's vast criminal _____ (justify) system, second chances are crucial.
3. He has _____ (acquire) a reputation as this country's premier solo violinist.
4. Migrating birds affect ecosystems both at home and at their winter destinations, and _____ (disrupt) the traditional routes could have unexpected side effects.
5. Have you had your child _____ (vaccinate) against whooping cough?
6. Development of renewable energy is promoted to achieve _____ (sustainable).
7. All government agencies must keep the alarm about the unexpected natural disasters, and take _____ (emerge) action and report.
8. The article II of the constitution of the USA _____ (prescription) the method of electing a president.
9. The Chinese nation has reached a point where its very existence is _____ (at stake) in the face of Japanese massive invasion of China.
10. The related data demonstrated that at present China has quite a high proportion of software _____ (pirate) rate.

Ex.5.4. Translate the following Chinese sentences into English.
1. 正面来看，互联网使人际交往更加便捷，且成本更低。(on the positive side)

   _____.

2. 任何借口都不能证明一国干涉别国内政是正当的。(justify)

   _____.

3. 香港将致力于提高内部和在国际的竞争力，推动经济持续增长。(sustain)

   _____.

4. 新的软件更新可能会提高计算机操作的效率。(be likely to)

   _____.

5. 计算机科学技术的快速发展对我们在数字时代的生活和工作方式产生了重大影响。(impact)

   _____.

Theme Writing

Directions: You are required to write a passage of about 120 words based on your understanding of the short passage below. Your writing should respond to the following questions:

1) What Apps are included in your mobile phone? List some of them.
2) Which App is the most important and beneficial to you? Why?

Passage:

There are three basic types of mobile apps if we categorize them by the technology used to code them:

1. Native apps are created for one specific platform or operating system. They are built specifically for a mobile device's operating system (OS). Thus, you can have native Android mobile apps or native iOS apps, not to mention all the other platforms and devices. Because they're built for just one platform, you cannot mix and match – say, use a Blackberry app on an Android phone or use an iOS app on a Windows phone.

2. Web apps are responsive versions of websites that can work on any mobile device or OS because they're delivered using a mobile browser. They behave similarly to native apps but are accessed via a web browser on your mobile device. They're not standalone apps in the sense of having to download and install code into your device. They're actually responsive websites that adapt its user interface to the device the user is on. In fact, when you come across the option to "install" a web app, it often simply bookmarks the website URL on your device.

3. Hybrid apps are combinations of both native and web apps, but wrapped within a native app, giving it the ability to have its own icon or be downloaded from an app store. They might have a home screen app icon, responsive design, fast performance, even be able to function offline, but they're really web apps made to look native.

第二章 拓展阅读

Reading 1

Digital Economy Expedites Quality Development

1   The world is accelerating into the age of digital economy. The World Bank predicts that

the share of the digital economy in the world's GDP will increase to 25 percent in 2025. The digital economy has apparently become the key driver for the world economy's recovery, and a growing number of countries are using it to boost their economic competitiveness.

2 The digital economy is booming in China too. New technologies such as big data, artificial intelligence and cloud computing are playing a bigger role in various fields and economic activities. In particular, with the information and communications technology marching into the 5G era, China, from being a follower, has become a leader in the field of digital technology.

3 By the end of May, China had the world's largest cyber infrastructures. All its prefecture-level cities have optic cable networks, more than 50 million Chinese people are connected through a 1000M fiber optic network, and about 1.7 million 5G base stations have been established in the country.

4 Benefitting from a series of favorable government policies and flexible regulations, such massive and advanced digital infrastructure is turning China into a leading global player in the digital economy. As the scale of the digital economy continues to expand, the list of innovations keeps increasing and its contribution to the country's GDP continues to grow.

5 Data show that from 2012 to 2021, the worth of China's digital economy has grown from 11 trillion yuan ($1.51 trillion) to more than 45 trillion yuan, and its weight in GDP has increased from 21.6 percent to 39.8 percent. In fact, the digital economy has become a core force driving China's high-quality economic development.

6 Also, during the 14th Five-Year Plan (2021—2025) period, government departments are expected to further ramp up efforts to overcome the technological barriers and to expedite the industrialization of the digital economy, in order to unleash its full potential for growth.

7 The booming digital economy, which is uniquely positioned to improve the speed of transmitting information, reduce the cost of trading and data processing, and more precisely allocate resources, will boost regional economic growth. Compared with traditional industries that depend on natural resources and are restricted by geography, the digital economy can cover much wider areas, providing many less-developed places an opportunity for boosting their local economy.

8 The digital economy can also help narrow regional disparity and promote common prosperity, especially during pandemics and other disease outbreaks when people have to maintain social distancing and sometimes stay at and work from home, dealing a heavy blow to the transportation, tourism and retail services. Besides, it can help integrate the real economy and the Internet Plus, and advance digitization across the country.

9 The digital economy has given birth to some new industries such as fresh food delivery, remote working, live-streaming, online medical care, which have facilitated high-quality economic and social development. And these emerging industries have prompted the government to increase investment in digital infrastructure and technologies such as 5G, AI and industrial internet.

10 Moreover, the digital technology has played a key role in the prevention and control of the pandemic, and the resumption of work and production, by effectively monitoring and analyzing the novel coronavirus, tracing cases, and helping infected people get medical treatment.

11 The digital economy can also promote the orderly flow and efficient allocation of resources across departments and regions in a relatively short period of time, even during emergencies, and become a stabilizing factor, cushioning the blow of pandemics and other emergencies.

12 But since most of China's studies on the digital economy remain stuck in qualitative analysis and policy proposals, the government needs to step up efforts to work out proper policies for the digital economy. And all industries and regions have to adapt to the "dual circulation" development paradigm to accelerate their high-quality development. Furthermore, it is necessary to theoretically explain and analyze new forms and models of the digital economy, explore policy measures to prevent the disorderly expansion of capital, and evaluate the impact of the government policies on the digital economy, so as to promote the digital transformation of the economy and speed up high-quality development in the new era.

China Daily, 2022-11-11

Comprehension Check for Reading 1

Direction: Skim the article and answer the following questions.

1. Which of the following facts can't prove the development of digital economy in the world?
    A) The share of the digital economy in the world's GDP will increase to 25 percent in a few years.
    B) Many countries are to increase their economic competitiveness with its help.
    C) The digital economy has become the only factor for the world economy's recovery.
    D) It is playing an increasingly important role in Chinese economic development.

2. What are the possible reasons for the improvement of Chinese digital infrastructure?
    A) Favorable government policies.   B) Innovation.
    C) Flexible regulations.   D) A&C.

3. Which of the following is not true about Chinese digital economy?
    A) It has developed more than four times in the past ten years.
    B) It took up almost 40% of Chinese GDP in 2021.
    C) Chinese government departments would spare no efforts to boost the industrial development during the 14th Five-Year Plan.

D) The digital economy can play a part in regional development without being limited by space.

4. Why does the author insist that during pandemics the digital economy is useful to lessen regional inequality?

A) Because at that time people have to communicate and work at home with its help while maintaining social distancing.

B) Because the pandemics will bring a heavy blow to the transportation, tourism and retail services.

C) Because digital economy provides many poorer places an opportunity in local economic development.

D) Because it is beneficial for both the development of real economy and the Internet Plus.

5. What's the author's attitude towards digital economy in China?

A) Uncertain.　　　　B) Hopeful.　　　　C) Neutral.　　　　D) Doubtful.

Reading 2

Cloudification Will Mean Upheaval in Telecoms

It will allow startups to challenge incumbent operators

1　In the computing clouds, startups can set up new servers or acquire data storage with only a credit card and a few clicks of a mouse. Now imagine a world in which they could as quickly weave their own wireless network, perhaps to give users of a fleet of self-driving cars more bandwidth or to connect wireless sensors.

2　As improbable as it sounds, this is the logical endpoint of a development that is picking up speed in the telecoms world. Networks are becoming as flexible as computing clouds: they are being turned into software and can be dialed up and down as needed. Such "cloudification", as it is known, will probably create as much upheaval in the telecoms industry as it has done in information technology (IT).

3　IT and telecoms differ in important respects. One is largely unregulated, the other overseen closely by government. Computing capacity is theoretically unlimited, unlike radio spectrum, which is hard to use efficiently. And telecoms networks are more deeply linked to the physical world. "You cannot turn radio towers into software," says Bengt Nordstrom of Northstream, a consultancy.

4　The data centers of big cloud-computing providers are packed with thousands of cheap servers, powered by standard processors. Telecoms networks, by contrast, are a collection of hundreds of different types of computers with specialized chips, each in charge of a different function, from text messaging to controlling antennae. It takes months, if not years, to set up a new service, let alone a new network.

5   But powerful forces are pushing for change. On the technical side, the current way of building networks will hit a wall as traffic continues to grow rapidly. The next generation of wireless technologies, called 5G, requires more flexible networks. Yet the most important factor behind cloudification is economic, says Stéphane Téral of IHS Markit, a market-research firm. Mobile operators badly need to cut costs, as the smartphone boom ends in many places and prices of mobile-service plans fall. The shift was evident at the Mobile World Congress in Barcelona in February. Equipment-makers' booths were plastered with diagrams depicting new technologies called NFV and SDN, which stand for "network-functions virtualization" and "software-defined networks". They turn specialized telecoms gear into software in a process called "virtualization".

6   Many networks have already been virtualized at their "core", the central high-capacity gear. But this is also starting to happen at the edges of networks — the antennae of a mobile network. These usually plug directly into nearby computers that control the radio signal. But some operators, such as SK Telecom in South Korea, have begun consolidating these "baseband units" in a central data centre. Alex Choi, SK Telecom's chief technology officer, wants "radio" to become the fourth component of cloud computing, after computing, storage and networking.

Spin me up, AT&T

7   The carrier that has pushed cloudification furthest is AT&T, America's largest operator. By the end of 2017 it wants to have more than half of its network virtualized. In areas where it has already upgraded its systems, it can now add to the network simply by downloading a piece of software. "Instead of sending a technician, we can just spin up a virtual machine," says Andre Fuetsch, AT&T's chief technology officer.

8   Even more surprising for a firm with a reputation for caution, AT&T has released the program that manages the newly virtualized parts of its network as open-source software: the underlying recipe is now available free. If widely adopted, it will allow network operators to use cheaper off-the-shelf gear — much as the rise of Linux, an open-source operating system, led to the commodification of hardware in data centers a decade ago.

9   If equipment-makers are worried about all this, they are not letting it show. Many parts of a network will not get virtualized, argues Marcus Weldon, chief technology officer of Nokia. And there will always be a need for specialized hardware, such as processors able to handle data packets at ever faster speeds. Still, Nokia and other telecoms-gear-makers will have to adapt. They will make less money from hardware and related maintenance services, which currently

form a big chunk of their revenues. At the same time, they will have to beef up their software business.

10   Cloudification may also create an opening for newcomers. Both Affirmed Networks and Mavenir, two American firms, for instance, are developing software to run networks on off-the-shelf servers. Mavenir wants to work with underdog operators "to bring the incumbents down", says Pardeep Kohli, its chief executive. If the history of cloud computing is any guide, the telecoms world may also see the rise of new players in the mould of Amazon Web Services (AWS), the e-commerce giant's fast-growing cloud-computing arm.

11   According to John Delaney of IDC, a research firm, the big barrier to cloudification is likely to be spectrum, which newcomers will still have to buy. But a clever entrepreneur may find ways to combine assets — unlicensed spectrum, fibre networks, computing power — to provide cheap mobile connectivity. Startups such as FreedomPop and Republic Wireless already offer "Wi-Fi first" mobile services, which send calls and data via Wi-Fi hotspots, using the mobile network as backup.

12   As the case of AWS shows, a potential Amazon Telecoms Services does not have to spring from the telecoms world. Amazon itself is a candidate. But carmakers, operators of power grids and internet giants such as Facebook could have a go: they are huge consumers of connectivity and have built networks. Facebook, for instance, is behind the Telecom Infra Project, another effort to open the network infrastructure. However things shake out, expect the telecoms world to become much more fluid in the coming years, just like IT before it.

The Economist 2017-4-12

Comprehension Check for Reading 2

Direction: Skim the article and answer the following questions.

1. What can be the logical endpoint of the development happening in the telecoms industry?

2. Are the data centers of big cloud-computing providers identical with that of the telecoms network?

3. What contributes cloudification most?

4. What would be the possible future confronted by the telecoms equipment-makers?

5. What will probably hinder the development of cloudification?

Unit Six
Artificial Intelligence

第六单元
人工智能技术

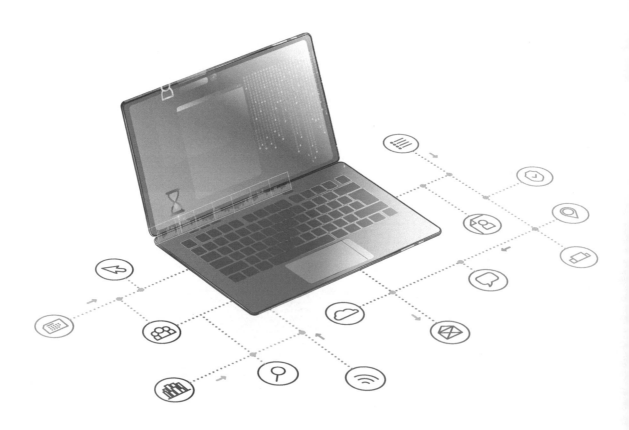

第一章　主题阅读

> **Text**

AI's New Frontier

1　Picture a computer that could finish your sentences, using a better turn of phrase; or use a snatch of melody to compose music that sounds as if you wrote it (though you never would have); or solve a problem by creating hundreds of lines of computer code — leaving you to focus on something even harder.

2　In a sense, that computer is merely the descendant of the power looms and steam engines that hastened the Industrial Revolution. But it also belongs to a new class of machine, because it grasps the symbols in language, music and programming and uses them in ways that seem creative. A bit like a human. The "foundation models" that can do these things represent a breakthrough in artificial intelligence, or AI. They, too, promise a revolution, but this one will affect the high-status brain work that the Industrial Revolution never touched. There are no guarantees about what lies ahead — after all, AI has stumbled in the past. But it is time to look at the promise and dangers of the next big thing in machine intelligence.

3　Foundation models are the latest twist on "deep learning" (DL), a technique that rose to prominence ten years ago and now dominates the field of AI. Loosely based on the networked structure of neurons in the human brain, DL systems are "trained" using millions or billions of examples of texts, images or sound clips. In recent years the ballooning cost, in time and money, of training ever-larger DL systems had prompted worries that the technique was reaching its limits. Some worried about an "AI winter". But foundation models show that building ever-larger and more complex DL does indeed continue to unlock ever more impressive new capabilities. Nobody knows where the limit lies.

4　The resulting models are a new form of creative, non-human intelligence. The systems are so sophisticated enough both to possess a grasp of language and also to break the rules coherently. A dog cannot laugh at a joke in the *New Yorker*, but an AI can explain why it is funny — a feat that is, frankly, sometimes beyond readers of the *New Yorker*. When we asked one of these models to create a collage using the title of this leader and nothing more, it came up with the cover art for our *American and Asian* editions (we tried to distract our anxious human designers with a different cover in our European editions).

5　Foundation models have some surprising and useful properties. The most strange and frightening of these is their "emergent" behaviour — that is, skills (such as the ability to get a joke or match a situation and a proverb) which arise from the size and depth of the models, rather than being the result of deliberate design. Just as a rapid succession of still photographs gives the sensation of movement, so trillions of binary computational decisions fuse into a simulacrum of fluid human comprehension and creativity that, whatever the philosophers may say, looks a lot like the real thing. Even the creators of these systems are surprised at their power.

6　This intelligence is broad and adaptable. True, foundation models are capable of behaving like an idiot, but then humans are, too. If you ask one who won the Nobel prize for physics in 1625, it may suggest Galileo, Bacon or Kepler, not understanding that the first prize was awarded in 1901. However, they are also adaptable in ways that earlier AIs were not, perhaps because at some level there is a similarity between the rules for manipulating symbols in disciplines as different as drawing, creative writing and computer programming. This breadth means that foundation models could be used in lots of applications, from helping find new drugs using predictions about how proteins fold in three dimensions, to selecting interesting charts from datasets and dealing with open-ended questions by trawling huge databases to formulate answers that open up new areas of inquiry.

7　That is exciting, and promises to bring great benefits, most of which still have to be imagined. But it also stirs up worries. Inevitably, people fear that AIs creative enough to surprise their creators could become harmful. In fact, foundation models are light-years from the sentient killer-robots beloved by Hollywood. Terminators tend to be focused, obsessive and blind to the broader consequences of their actions. Foundational AI, by contrast, is fuzzy. Similarly, people are anxious about the large amounts of power training these models consume and the emissions they produce. However, AIs are becoming more efficient, and their insights may well be essential in developing the technology that accelerates a shift to renewable energy.

8　A more penetrating worry is over who controls foundation models. Training a really large system such as Google's paLM costs more than $10 million and requires access to huge amounts of data — the more computing power and the more data the better. This raises the worries of a technology concentrated in the hands of a small number of tech companies or governments. If so, the training data could further deepen the world's biases — and in a particularly oppressive and unpleasant way.

9　Would you trust a ten-year-old whose entire sense of reality had been formed by surfing the internet? Might Chinese-and American-trained AIs be recruited to an ideological struggle to bend minds? What will happen to cultures that are poorly represented online?

10　And then there is the question of access. For the moment, the biggest models are restricted, to prevent them from being used for immoral purposes such as generating fake news stories. OpenAI, a startup, has designed its model, called dall-e 2, in an attempt to stop it producing violent or erotic images. Firms are right to fear abuse, but the more powerful these models are, the more limiting access to them creates a new elite. Self-regulation is unlikely to resolve the dilemma.

Bring on the Revolution

11　For years it has been said that AI-powered automation poses a threat to people in repetitive, routine jobs, and that artists, writers and programmers were safer. Foundation models challenge that assumption. But they also show how AI can be used as a software sidekick to enhance productivity. This machine intelligence does not resemble the humankind, but offers something entirely different. Handled well, it is more likely to complement humanity than take control of it.

The Economist 2022-06

Vocabulary

anxious *adj.* 焦虑的，担心的；渴望的，急切的 (...to/about *sth.* ; ...that...)
abuse *vt.&n.* 虐待；滥用；辱骂
assumption *n.* 假定，假设
accelerate *v.* (使) 加快，促进；(车辆或驾驶者) 加速
automation *n.* 自动化
adaptable *adj.* 能适应的；可改变 (以适应新用途) 的
balloon *n.* 气球；*vi.* 激增，膨胀；*adj.* 像气球般鼓起的
binary *adj. & n.* 二进制 (的)
breadth *n.* 宽度，幅度；广度
bend *vi.* 弯曲；*adj.* 弯曲的；*n.* 弯道
compose *v.* 构成；作曲 (词)
coherently *adv.* 连贯地
collage *n.& vt.* (创作) 抽象拼贴画
complement *v.* 补充，补足；*n.* 补足物，衬托物
descendant *n.* 后裔，子孙；衍生物；*adj.* 下降的；祖传的
deliberate *adj.* 故意的；小心翼翼的；深思熟虑的；*v.* 仔细考虑，认真商讨

discipline *n.* 纪律；处罚；训练；自律；(尤指大学的) 科目；*v.* 惩罚；训练
dimension *n.* 大小，尺寸；(空间的) 维度；规模；方面
dilemma *n.* (进退两难的) 窘境，困境
emergent *adj.* 新兴的，处于发展初期的；(特征) 突现的
emission *n.* 排放物，散发物；(尤指光、热、气等的) 散发，排放
erotic *adj.* 色情的
fuzzy *adj.* 模糊的；含混的；迷糊的
fuse *n.* 保险丝，熔丝；导火线；*v.* 熔合，结合
formulate *v.* 制定；认真阐述
hasten *vt.* 加速；急忙进行，赶紧说 (或做)
inquiry *n.* 询问；调查
inevitably *adv.* 不可避免地
ideological *adj.* 思想体系的，意识形态的
immoral *adj.* 不道德的，邪恶的
loom *vi.* (常以可怕的方式) 赫然出现；隐约显现
loosely *adv.* 宽松地；放荡地；轻率地
manipulate *vt. & n.* 控制；操纵

neuron *n.* 神经元，神经单位
obsessive *adj.* 着迷的，迷恋的
oppressive *adj.* 压迫的；压抑的
prominence *n.* 重要性，著名
generate *v.* 产生，引起
protein *n. & adj.* 蛋白质 (的)
penetrating *adj.* 渗透的；有洞察力的
repetitive *adj.* 重复乏味的；多次重复的
recruit *vt.* 招收 (新成员)；招募 (新兵)

n. 新兵；新成员
resemble *vt.* 像，与……相似
snatch *vt.* 夺取；*n.* 片段
stumble *vt.* 跟跄；绊脚
sensation *n.* 感觉；轰动
sophisticated *adj.* 老练的；高级的，复杂的
simulacrum *n.* 像；幻影；影
sidekick *n.* 伙伴，老朋友
trillion [数]万亿

Cultural Notes

Deep Learning (DL)　The deep learning is a new research direction in the field of Machine Learning (ML). It is introduced into machine learning to make it closer to the original target — Artificial Intelligence (AI). Deep learning is the learning of the internal rules and representation layers of sample data. The information acquired during the learning process is of great help to the interpretation of data such as text, images and sounds. Its ultimate goal is to give machines the analytical learning capabilities of humans, able to recognize data such as words, images and sounds. Deep learning is a complex machine learning algorithm that has achieved far greater results in speech and image recognition than previous techniques.

The New Yorker　*The New Yorker*, is a comprehensive American intellectual, literary and art magazine. It focuses on nonfiction, including reporting and commentary on politics, international affairs, popular culture and art, science and technology, and business. It also publishes literary works, mainly short stories and poetry, as well as humorous sketches and cartoons.

Foundation Model　The foundation model is the AI grand model, a model that can be adapted to a series of downstream tasks by training on a large scale and wide range of data. The foundation model is a milestone technology for AI to move towards general intelligence. As a landmark technology of the new generation of artificial intelligence, deep learning relies entirely on models to automatically learn knowledge from data. While significantly improving performance, it also faces the contradiction of surging general data and lacking special data. AI grand model has two attributes of "large-scale" and "pre-training". Pre-training on massive general data is required before practical task modeling, which can greatly improve the generalization, universality and practicability of AI.

Terminator   It's an intelligent robot from the 1984 science fiction film *Terminator*. The movie is about a future world where robots are already controlling the world. The robots want to completely take over the world and kill the human race, but they encounter the tenacious resistance of the human elite Connor. The Terminator T-800 (Arnold Schwarzenegger) is sent back to 1984 to kill Connor's mother, Sarah, in order to kill Connor's birth.

Google's paLM   On April 4, 2022, Google announced its Pathways language model (PaLM). With 540 billion parameters, PaLM continues the broader technology trend of building larger language models. PaLM is just a little bigger than Microsoft/Nvidia's Megataton-Turing NLG, almost twice as big as DeepMind's Gopher, and much bigger than Open AI's GPT-3 (175 billion parameters). Large language models (LLMs) have shown what they can do for a wide variety of tasks.

OpenAI   In 2015, OpenAI was founded by Musk, Altman, president of US startup incubator Y Combinator, and Peter Thiel, co-founder of global online payment platform PayPal and other Silicon Valley tech tycoons. On June 21, 2016, OpenAI announced its main goals, including building "universal" bots and chatbots that use natural language. On July 22, 2019, Microsoft invested $1 billion in OpenAI, and the two sides will work together to develop artificial intelligence technology for Azure cloud platform services. In June 2022, quantum computing expert and ACM Computing Award winner Scott Aaronson announced that he would join the company.

Galileo   Galileo Galilei (15 February 1564 — 8 January 1642), was an Italian astronomer, physicist, engineer and founder of modern natural science in Europe. Galileo has been called "the father of observational astronomy", "the father of modern physics", "the father of the scientific method", "the father of modern science".

Bacon   Roger Bacon (1214 — 1293) was an English philosopher and natural scientist with a materialistic tendency, a famous nominalist, and a pioneer of experimental science. With a wide range of knowledge, he was known as the "Strange Doctor".

Kepler   Johannes Kepler is a German astronomer, mathematician and astrologer. Kepler discovered three laws of planetary motion, namely the law of orbit, the law of area and the law of period. These three laws eventually earned him the name "Sky Lawgiver". At the same time, he also made important contributions to optics and mathematics, he is the founder of modern experimental optics.

Read between Lines

Direction: Read the passage carefully and answer the questions below.

Question 1: In what way is the computer like a human? (Para.1~2)

_____.

Question 2: What does the "AI winter" refer to? (Para.3~4)

_____.

Question 3: What is the most strange and frightening property of foundation models? (Para.5~6)

_____.

Question 4: How many kinds of worries are mentioned in Para 7 & 8? What are they? (Para7~8)

_____.

Question 5: Why does the author mention a ten-year-old? (Para.9~10)

_____.

Question 6: What does the sentence "Foundation models challenge that assumption." imply? (Para.11)

_____.

Words and Expressions

Ex. 6.1. *Write the word according to the definition. The first letter is given.*

1. d_____: the practice of making people obey rules or standards of behavior, and punishing them when they do not.
2. a_____: able to change ideas or behavior in order to deal with new situations.
3. i_____: a question you ask in order to get some information.
4. f_____: invent a plan or proposal, thinking about the details carefully.
5. s_____: machine, device, or method is more advanced or complex than others.
6. c_____: form the substance of; write music or produce a literary work.

Ex. 6.2. *Fill in the blanks of the appropriate word from the box below. Change the form if necessary.*

| hasten | anxious | accelerate | immoral | stir up |
| obsessive | inevitably | compose | adaptable | deliberate |

1. The thieves must have _____ triggered the alarm and hidden inside the house.
2. We watched him go out on a run the first time and wondered how long this _____ would last.
3. Most software programs allow you to _____ e-mails offline.
4. The foreign minister admitted he was still _____ about the situation in the country.
5. It is_____ that most senior students have to choose between pursuing their studies and see king employment.
6. Emphasis was placed on the school as a transmitter of _____ values.
7. The world will be different, and we will have to be prepared to _____ to the change.
8. To _____ the development of science from the lab to the market place, Wuhan University is investing money in our good ideas.
9. If you keep your mouth shut, you are unlikely to _____ controversy.
10. It is reported that more efforts will be made in the years ahead to _____ the agricultural reform.

Ex. 6.3. *Complete each sentence below with the appropriate form of word in bracket.*

1. He appeared hardly capable of conducting a _____ (coherently) conversation.
2. The story was given a _____ (prominence) position on the front page.
3. The cultural life of the country will sink into recession unless more writers and artists _____ (emergent).
4. I create a new instance of the type for every _____ (manipulate).
5. It is impossible to _____ (prediction) what the eventual outcome will be.
6. The government has pledged to clean up industrial _____ (emission).
7. X-rays can _____ (penetrating) many objects.
8. Brainstorming is a good way of _____ (generate) ideas.
9. You _____ (assumption) his innocence before hearing the evidence against him.
10. She _____ (repetitive) her call yesterday for an investigation into the incident.

Ex. 6.4. *Translate the following Chinese sentences into English.*

1. 在计算机科学技术领域，严格的学科纪律对于确保研究成果的可靠性和推动技术创新至关重要。(discipline)

2. 你知道怎么在电脑上作曲吗？(compose)

_____.

3. 在过去的一年中，开发团队渴望发布新的软件版本。(anxious)

_____.

4. 在信息技术行业，对于新技术和新应用的持续调查是推动行业进步和创新的关键。(inquiry)

_____.

5. 一批专家声称，他们可以通过一种电脑程序预测房价。(predict)

_____.

Theme Writing

Directions: You are required to write a passage of about 120 words based on your understanding of the short passage below. Your writing should respond to the following questions:

1) Do you think artificial intelligence is a bless or a curse to human beings?
2) How can human beings make better use of the artificial intelligence?

Passage:

  Can machines really think? The artificial intelligence, such as a computer that thinks like a human being is scary. Is building a machine that thinks like a human really possible? We are ever closer to building an AI that thinks like a human. When it comes to this issues, different people offer different views, some people think that machine has feelings like human beings is interesting and it may be a better server to human; while the other think it is dangers, it may causes a revolt.

  People who approved of human feelings machine think that once robot has specific feelings, such as happy, sad, anger, they might be more humanize. For example, maybe in the future a robot nanny will replace a real human nanny, who are work more effective and without any complain. If they have real emotion, they are more perfect, and more like a company but not a cool machine. People who against human robot argue that once the robot is more intelligent than we think, that maybe a great tribulation to human beings. There has a potential risks that once the robot is smart enough, they may unwilling to be human's server anymore, they may want to be legally citizens, or even worse, to be the owner of the world. It is possible because they are smart and they are stronger compare with human beings. It is not sure what will happen in the future, having robot to serve for human beings is a good thing, but the issue of artificial intelligence is still controversial.

第二章 拓展阅读

Reading 1

AI's Role in Driving Growth Bigger

Companies integrating cutting-edge tech show competitive advantages

1 Chinese enterprises are planning to increase investment in artificial intelligence as AI is increasingly perceived as one of the drivers in revenue growth and business transformation and integrated into the real economy, according to a new report from global consultancy Accenture.

2 The report said 34 percent of surveyed Chinese companies used more than 30 percent of their tech budgets for AI projects in 2021. By 2024, the percentage of companies investing over 30 percent of their tech budgets in AI is expected to increase to 64 percent.

3 AI has become an important competitive advantage for enterprises. Research has found that 13 percent of surveyed Chinese enterprises have used AI to outpace their competitors. This group is dubbed "AI Achievers", with a score of 64 on the maturity scale, almost double that of others and correlating with 50 percent higher revenue growth than their peers.

4 AI maturity is the degree to which companies outperform their peers in a combination of AI-related foundational and differentiating capabilities, which include cloud computing, data and AI algorithms, as well as AI strategy, talent and innovation culture, Accenture said.

5 The report estimated the proportion of "AI Achievers" will increase rapidly to 34 percent by 2024. In addition, more than half of the surveyed Chinese enterprises are "AI Experimenters", who have barely scratched the surface of AI's potential and lack mature AI strategies and related capabilities.

6 Although some industries like high-tech are currently far ahead in their AI maturity, the gap will likely narrow considerably by 2024. There is enormous room for growth in AI adoption across all industries and opportunities for those companies that choose to seize it, the report noted.

7 For instance, the automotive sector is betting on a big surge in sales of AI-powered self-driving vehicles, while aerospace firms anticipate continued demand for AI-enabled remote systems. The life sciences industry will expand its use of AI in efficient drug development.

8 The research surveyed 250 Chinese companies in 17 industries, including retail, telecommunications, chemicals, energy, financial services and healthcare between July and

September, with their sales revenue surpassing $1 billion in 2021.

9   "We believe every part of every business must be transformed by technology, data and AI, in some cases resulting in total enterprise reinvention," said Sanjeev Vohra, global lead for applied intelligence at Accenture.

10   "Adopting AI at scale and embedding it deeper in all aspects of business is no longer a choice but a necessity and opportunity facing every industry, organization and leader," Vohra added.

11   AI, a key technology for driving digital transformation, is playing an increasingly important role in accelerating China's push for industrial upgrading and promoting the in-depth integration of the digital economy and the real economy.

12   China has issued a plan setting benchmarks for its AI sector, with the value of core AI industries predicted to exceed 1 trillion yuan ($143.6 billion), making China one of the global leaders in such innovation by 2030.

13   According to the Ministry of Industry and Information Technology, the value of China's core AI industries exceeds 400 billion yuan and the number of related enterprises stands at more than 3,000, with breakthroughs being made in key core technologies such as smart chips and open-source frameworks.

14   An increasing number of Chinese companies will achieve digital transformation with the help of data and AI, said Chan Tzeh Chyi, managing director of strategy and consulting, applied intelligence lead and chief data scientist at Accenture Greater China.

15   He added that business executives should accelerate steps to promote the large-scale use of AI, prioritize long and short-term AI investment and allow AI to better integrate into enterprises' overall strategies.

16   Xiang Ligang, director-general of the Information Consumption Alliance, a telecom industry association, underscored the significance of developing digital technologies represented by AI, which will inject fresh impetus into the country's economic growth and speed up digital and intelligent upgrades in enterprises.

17   The in-depth integration of digital technologies with the real economy will further reinforce China's advantages in global supply chains, he said.

China Daily 2022-12-28

Comprehension Check for Reading 1

Direction: Skim the article and answer the following questions.

1. What is the main idea of this passage?

   A) There is enormous room for AI growth across all industries.

   B) AI is increasingly driving integration of digital technologies with the real economy.

C) The current development of AI among different industries in China.

D) High-tech industries are currently far ahead in the AI maturity.

2. Which of the following statement is TRUE according the passage?

A) The companies will increase their tech budgets in AI from 30% to 64% by 2024.

B) The number of AI Achievers is almost double those of other Chinese enterprises.

C) AI Experimenters have almost scratched the surface of AI's potential.

D) The gap in AI maturity among industries will likely narrow by 2024.

3. Which of the following is not included in AI maturity?

A) Talent.　　　B) Data.　　　C) AI market.　　　D) AI strategy.

4. What is the purpose of the example of automotive sector?

A) To prove that automotive sector is investing more in AI technology.

B) To prove that AI is better used in auto industry.

C) To illustrate the company will have opportunities if it adopts AI.

D) To illustrate the company's big development with AI.

5. How can Chinese enterprises better integrate AI into their business?

A) To undergo total enterprise reinvention.

B) To prioritize long and short-term AI investment.

C) To put all the investment in AI technology.

D) To pay more attention to research and development of AI products.

Reading 2

Could OpenAI be the Next Tech Giant?

What the business of AI's leading startup says about the technology's future

1　The creation of a new market is like the start of a long race. Competitors jockey for position as spectators excitedly clamour. Then, like races, markets enter a calmer second phase. The field orders itself into leaders and laggards. The crowds thin.

2　In the contest to dominate the future of artificial intelligence, OpenAI, a startup backed by Microsoft, established an early lead by launching ChatGPT last November. The app reached 100m users faster than any before it. Rivals scrambled. Google and its corporate parent, Alphabet, rushed the release of a rival chatbot, Bard. So did startups like Anthropic. Venture capitalists poured over $40bn into AI firms in the first half of 2023, nearly a quarter of all venture dollars this year. Then the frenzy died down. Public interest in AI peaked a couple of months ago, according to data from Google searches. Unique monthly visits to ChatGPT's website have declined from 210m in May to 180m now (see chart 1).

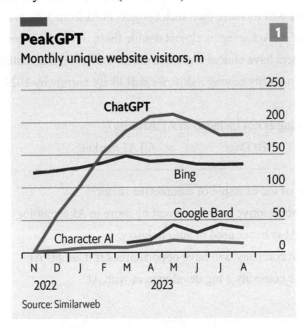

Chart 1　The Economist

3　The emerging order still sees OpenAI ahead technologically. Its latest AI model, GPT-4, is beating others on a variety of benchmarks (such as an ability to answer reading and maths questions). In head-to-head comparisons, it ranks roughly as far ahead of the current runner-up, Anthropic's Claude 2, as the world's top chess player does against his closest rival — a decent lead, even if not insurmountable. More important, OpenAI is beginning to make real money. According to The Information, an online technology publication, it is earning revenues at an annualized rate of $1bn, compared with a trifling $28m in the year before ChatGPT's launch.

4　Can OpenAI translate its early edge into an enduring advantage, and join the ranks of big tech? To do so it must avoid the fate of erstwhile tech pioneers, from Netscape to Myspace, which were overtaken by rivals that learnt from their early successes and stumbles. And as it is a first mover, the decisions it takes will also say much about the broader direction of a nascent industry.

5　OpenAI is a curious firm. It was founded in 2015 by a clutch of entrepreneurs including Sam Altman, its current boss, and Elon Musk, Tesla's technophilic chief executive, as a non-profit venture. Its aim was to build artificial general intelligence (AGI), which would equal or surpass human capacity in all types of intellectual tasks. The pursuit of something so outlandish meant that it had its pick of the world's most ambitious ai technologists. While working on an AI that could master a video game called "Dota", they alighted on a simple approach that involved harnessing oodles of computing power, says an early employee who has since left. When in 2017 researchers at Google published a paper describing a revolutionary machine-learning technique they christened the "transformer", OpenAI's boffins realised that they could scale it up by combining untold quantities of data scraped from the internet with processing oomph. The result was the general-purpose transformer, or GPT for short.

6   Obtaining the necessary resources required OpenAI to employ some engineering of the financial variety. In 2019 it created a "capped-profit company" within its non-profit structure. Initially, investors in this business could make 100 times their initial investment — but no more. Rather than distribute equity, the firm distributes claims on a share of future profits that come without ownership rights ("profit-participation units"). What is more, OpenAI says it may reinvest all profits until the board decides that OpenAI's goal of achieving AGI has been reached. OpenAI stresses that it is a "high-risk investment" and should be viewed as more akin to a "donation". "We're not for everybody," says Brad Lightcap, OpenAI's chief operating officer and its financial guru.

7   Maybe not, but with the exception of Mr Musk, who pulled out in 2018 and is now building his own AI model, just about everybody seems to want a piece of OpenAI regardless. Investors appear confident that they can achieve venture-scale returns if the firm keeps growing. In order to remain attractive to investors, the company itself has loosened the profit cap and switched to one based on the annual rate of return (though it will not confirm what the maximum rate is). Academic debates about the meaning of AGI aside, the profit units themselves can be sold on the market just like standard equities. The firm has already offered several opportunities for early employees to sell their units.

8   SoftBank, a risk-addled tech-investment house from Japan, is the latest to be seeking to place a big bet on OpenAI. The startup has so far raised a total of around $14bn. Most of it, perhaps $13bn, has come from Microsoft, whose Azure cloud division is also furnishing OpenAI with the computing power it needs. In exchange, the software titan will receive the lion's share of OpenAI's profits — if these are ever handed over. More important in the short term, it gets to license OpenAI's technology and offer this to its own corporate customers, which include most of the world's largest companies.

9   It is just as well that OpenAI is attracting deep-pocketed backers. For the firm needs an awful lot of capital to procure the data and computing power necessary to keep creating ever more intelligent models. Mr Altman has said that OpenAI could well end up being "the most capital-intensive startup in Silicon Valley history". OpenAI's most recent model, gpt-4, is estimated to have cost around $100m to train, several times more than GPT-3.

10   For the time being, investors appear happy to pour more money into the business. But

they eventually expect a return. And for its part OpenAI has realized that, if it is to achieve its mission, it must become like any other fledgling business and think hard about its costs and its revenues.

11 GPT-4 already exhibits a degree of cost-consciousness. For example, notes Dylan Patel of SemiAnalysis, a research firm, it was not a single giant model but a mixture of 16 smaller models. That makes it more difficult — and so costlier — to build than a monolithic model. But it is then cheaper to actually use the model once it has been trained. because not all the smaller models need be used to answer questions. Cost is also a big reason why OpenAI is not training its next big model, GPT-5. Instead, say sources familiar with the firm, it is building GPT-4.5, which would have "similar quality" to GPT-4 but cost "a lot less to run".

12 But it is on the revenue-generating side of business that OpenAI is most transformed, and where it has been most energetic of late. AI can create a lot of value long before AGI brains are as versatile as human ones, says Mr Lightcap. OpenAI's models are generalist, trained on a vast amount of data and capable of doing a variety of tasks. The ChatGPT craze has made OpenAI the default option for consumers, developers and businesses keen to embrace the technology. Despite the recent dip, ChatGPT still receives 60% of traffic to the top 50 generative-AI websites, according to a study by Andressen Horowitz, a venture-capital (vc) firm which has invested in OpenAI (see chart 2).

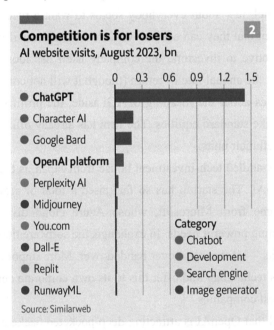

Chart 2 The Economist

13 Yet OpenAI is no longer only — or even primarily — about ChatGPT. It is increasingly becoming a business-to-business platform. It is creating bespoke products of its own for big corporate customers, which include Morgan Stanley, an investment bank. It also offers

tools for developers to build products using its models; on November 6th it is expected to unveil new ones at its first developer conference. And it has a $175m pot to invest in smaller AI startups building applications on top of its platform, which at once promotes its models and allows it to capture value if the application-builders strike gold. To further spread its technology, it is handing out perks to AI firms at Y Combinator, a Silicon Valley startup nursery that Mr Altman used to lead. John Luttig of Founders Fund (a VC firm which also has a stake in OpenAI), thinks that this vast and diverse distribution may be even more important than any technical advantage.

14 Being the first mover certainly plays in OpenAI's favour, GPT-like models' high fixed costs erect high barriers to entry for competitors. That in turn may make it easier for OpenAI to lock in corporate customers. If they are to share internal company data in order to fine-tune the model to their needs, many clients may prefer not to do so more than once — for cyber-security reasons, or simply because it is costly to move data from one AI provider to another, as it already is between computing clouds. Teaching big models to think also requires lots of tacit engineering know-how, from recognising high-quality data to knowing the tricks to quickly debug the source code. Mr Altman has speculated that fewer than 50 people in the world are at the true model-training frontier. A lot of them work for OpenAI.

15 These are all real advantages. But they do not guarantee OpenAI's continued dominance. For one thing, the sort of network effects where scale begets more scale, which have helped turn Alphabet, Amazon and Meta into quasi-monopolists in search, e-commerce and social networking, respectively, have yet to materialize. Despite its vast number of users, GPT-4 is hardly better today than it was six months ago. Although further tuning with user data has made it less likely to go off the rails, its overall performance has changed in unpredictable ways, in some cases for the worse.

16 Being a first mover in model-building may also bring some disadvantages. The biggest cost for modellers is not training but experimentation. Plenty of ideas went nowhere before the one that worked got to the training stage. That is why OpenAI is estimated to have lost $500m last year, even though GPT-4 cost one-fifth as much to train. News of ideas that do not pay off tends to spread quickly throughout AI world. This helps OpenAI's competitors avoid going down costly blind alleys.

17 As for customers, many are trying to reduce their dependence on OpenAI, fearful of being locked into its products and thus at its mercy. Anthropic, which was founded by defectors from OpenAI, has already become a popular second choice for many ai startups. Soon businesses may have more cutting-edge alternatives. Google is building Gemini, a model believed to be more powerful than GPT-4. Even Microsoft is, despite its partnership with OpenAI, something of a competitor. It has access to GPT-4's black box, as well as a vast sales force with long-standing ties to the world's biggest corporate IT departments. This array of choices diminishes OpenAI's pricing power. It is also forcing Mr Altman's firm to keep training better models if it wants to

stay ahead.

18   The fact that OpenAI's models are a black box also reduces its appeal to some potential users, including large businesses concerned about data privacy. They may prefer more transparent "open-source" models like Meta's llama 2. Sophisticated software firms, meanwhile, may want to build their own model from scratch, in order to exercise full control over its behaviour.

19   Others are moving away from generality — the ability to do many things rather than just one thing — by building cheaper models that are trained on narrower sets of data, or to do a specific task. A startup called Replit has trained one narrowly to write computer programs. It sits atop Databricks, an AI cloud platform which counts Nvidia, a $1trn maker of specialist AI semiconductors, among its investors. Another called Character AI has designed a model that lets people create virtual personalities based on real or imagined characters that can then converse with other users. It is the second-most popular AI app behind ChatGPT.

20   The core question, notes Kevin Kwok, a venture capitalist (who is not a backer of OpenAI), is how much value is derived from a model's generality. If not much, then the industry may be dominated by many specialist firms, like Replit or Character AI. If a lot, then big models such as those of OpenAI or Google may come out on top.

21   Mike Speiser of Sutter Hill Ventures (another non-OpenAI backer) suspects that the market will end up containing a handful of large generalist models, with a long tail of task-specific models. If AI turns out to be all it is cracked up to be, being an oligopolist could still earn OpenAI a pretty penny. And if its backers really do see any of that penny only after the company has created a human-like thinking machine, then all bets are off.

The Economist 2023-07-18

Comprehension Check for Reading 2

Direction: Skim the article and answer the following questions.

1. What did OpenAI's rivals do when it launched ChatGPT?

___

2. What's the first thing OpenAI should do to join the ranks of big tech?

___

3. How can OpenAI achieve its mission?

_____

4. Why is it favourable for OpenAI to be the first mover?

_____

5. What kind of businesses are not likely be OpenAI's potential customers?

_____

Unit Seven

Fifth Generation Mobile Communication Technology

—— 第七单元 ——

5G技术

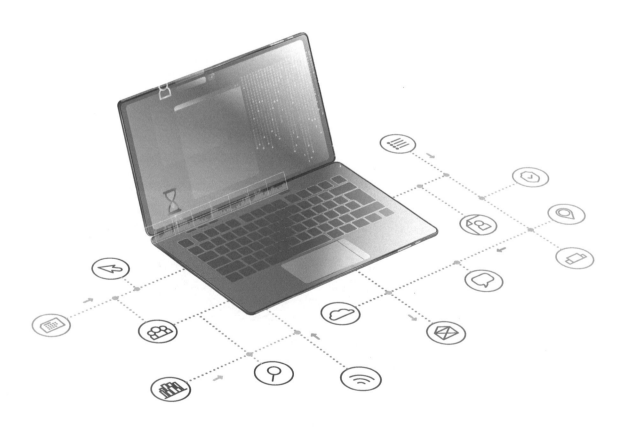

第一章 主题阅读

Text

How 5G Can Unlock the Potential of Smart Homes

1 5G technology has been one of the hottest topics in the tech world in recent years. Fifth-generation mobile network technology has been the subject of discussion everywhere from healthcare to sports to entertainment to education. However, one of the sectors where 5G technology has proven to have a real impact is the real estate industry.

2 In this article, we will explore how 5G technology is transforming the way properties are bought, sold and managed. We will look at the benefits of 5G technology for real estate professionals, homeowners, and buyers, and discuss how this technology is helping to create smarter, more efficient homes.

3 5G promises to significantly enhance speed, latency and capacity. In comparison to 4G networks, 5G networks can enable download rates up to 100 times faster, with latency as little as one millisecond. In light of this, 5G networks are more suited for using apps like virtual reality (VR) and streaming video than ever before since they can support more data and devices. Furthermore, 5G networks are intended to be more durable and dependable than earlier cellular technology generations, offering more constant service and fewer lost connections.

4 One of the key differences between 5G and previous generations of cellular technology is its ability to support a wider range of devices and use cases. 5G networks are designed to support everything from smartphones and tablets to connected cars and IOT devices. This means that 5G networks will be able to handle a much wider range of data types and traffic patterns than previous generations of cellular technology.

5 5G technology is transforming the way people interact with property. For example, 5G technology is enabling the use of VR and augmented reality (AR) in property viewing and preparation. This means that buyers and sellers can have a more immersive and realistic property viewing experience without having to physically be at the property.

6 5G technology is also enabling real estate professionals to collaborate more effectively and work remotely. Real estate agents and brokers can access real-time property data and information from anywhere with a stable internet connection.

7 5G technology also offers significant benefits for real estate professionals. Real estate

agents and brokers can take advantage of increased connectivity and data speeds to collaborate more effectively and work faster. The ability to access property information in real time, regardless of location, allows real estate professionals to provide faster and more efficient service to their clients.

8 5G technology also allows real estate professionals to collaborate with other professionals, such as lawyers and appraisers, more easily. The ability to share data and collaborate in real time is a major change for the real estate industry and has been welcomed by many professionals.

9 5G technology is transforming the way people interact with their homes. With faster download speeds and connectivity, 5G technology is enabling more advanced smart homes. 5G technology is making it possible to integrate more smart devices into the home, such as smart thermostats, security systems and surveillance cameras, and connected home appliances. This allows homes to become more efficient and automated, which in turn reduces energy costs and improves homeowners' safety and comfort.

10 For the real estate industry, this means that there will be a plethora of extra amenities and technologies that will be offered alongside the home itself once 5G makes its breakthrough. For example, as mentioned before, this technology can allow for more sophisticated surveillance systems in the properties, which can give real estate agents the ability to show the property without worrying about anything happening to it and being able to monitor it the whole time.

11 5G technology also has the potential to revolutionize the way commercial buildings are managed. Commercial buildings can use 5G technology to control and manage their security, power and heating systems remotely. In addition, 5G technology can also improve the energy efficiency of commercial buildings by enabling better management and control of energy consumption.

12 While 5G technology has great potential to transform the real estate industry, there are still challenges and hurdles to overcome before widespread adoption can be achieved. One of the biggest challenges is the need for additional infrastructure to support 5G technology. Additional communications towers and more advanced network infrastructure are needed to enable high-quality 5G connectivity. This can be costly and require significant investment by Telecom service providers.

13 Another major obstacle is the lack of awareness and education about 5G technology. Many real estate professionals and homeowners may not be familiar with the benefits of 5G technology and how it can improve their operations and experiences. More effort is needed to educate consumers about 5G technology and how it can benefit them.

14 In summary, 5G technology has the potential to revolutionize the way properties are bought, sold and managed in the real estate industry. From creating more advanced smart homes to improving the energy efficiency of commercial buildings, 5G technology is driving innovation and improving the customer experience across the industry. While there are still challenges and

hurdles to overcome, it is clear that 5G technology has the potential to completely transform the way we interact with each other.

Forbes 2023-05-10

Vocabulary

advanced *adj.* 先进的；高级的，高等的；晚期的
agent *n.* 代理人，代理商
amenity *n.* 方便条件，便利设施；舒适，惬意；礼仪
appraiser *n.* 鉴定人，评价人；估价官
augmented *adj.* 增广的；增音的；扩张的
automated *adj.* 自动化的
broker *n.* 经纪人，中间人；*v.* 协调，安排
capacity *n.* 能力，才能；容积；*adj.* 无虚席的，满场的
cellular *adj.* 由细胞组成的，细胞的；(电话系统) 蜂窝状的，蜂窝式的；*n.* 移动电话
client *n.* 客户，委托人；客户机，用户端
collaborate *v.* 合作，协作；勾结，通敌
commercial *adj.* 商业的，商务的；商业化的
connectivity *n.* [数] 连通性；连通 (性)，联结 (度)
constant *adj.* 持续不断的，经常发生的；*n.* 常数，恒量；不变的事物
consumption *n.* 消费，消耗；食用，引用；吸入；专用
dependable *adj.* 可靠的，可信赖的
download *v.* 下载；*n.* 下载，下载的文件或程序
durable *adj.* 持久的，耐用的；*n.* 耐用品
efficient *adj.* 生效的，效率高的；(人) 有能力的，能胜任的

estate *n.* 财产，遗产；大片私有土地
familiar *adj.* 熟悉的；常见的，普通的；非正式的，随和的
healthcare *n.* 医疗保健服务
homeowner *n.* 私房屋主；自己拥有住房者
hurdle *n.* 障碍，难关；*v.* (奔跑中) 跳越 (某物)；克服(障碍或困难)
immersive *adj.* 身临其境的；(计算机系统或图像) 沉浸式虚拟现实的
infrastructure *n.* 基础设施，基础建设；下面结构；永久性军事设施
innovation *n.* 新事物，新方法；革新，创新
interact *v.* 相互交流，互动；相互作用，相互影响
latency *n.* 延迟，时延；潜伏；潜在因素
millisecond *n.* 毫秒，千分之一秒
monitor *n.* 显示器，监控器；*v.* 监视；监听(外国广播或电话)
potential *adj.* 潜在的，可能的 *n.*(事物的) 潜力，可能性；(人的) 潜能
previous *adj.* 以前的，先前的；(时间或顺序上) 稍前的
property *n.* 所有物，财产；地产，房地产
revolutionize *v.* 使发生革命性巨变，使彻底变革；发动革命
sector *n.* 区域，部分；(尤指商业、贸易等的) 部门，行业
significantly *adv.* 显著地，相当数量地；值得注意的是
streaming *n.* 串流，流式传播；*adj.* (计算

机) 流式传输的；v.(人，东西) 流动，涌动
surveillance n. 监视，监察
thermostat n. 温度自动调节器，恒温器；v. 为……配备恒温器，用恒温器控制

transform v. 使改观，使变形，使转化；变换(电流) 的电压
unlock v. 开……的锁；释放 (潜能)，揭开 (秘密)；开启

Cultural Notes

Virtual Reality (VR)　　Virtual reality technology, also known as virtual reality or spiritual realm technology. It is a new practical technology developed in the 20th century. Virtual reality technology concludes computer, electronic information, simulation technology and its basic realization is based on computer technology by using and combining three-dimensional graphics technology, multimedia technology, simulation technology, display technology, servo technology and other high-tech latest development results. With the help of computers and other equipment, it aims to produce a realistic three-dimensional visual, tactile, olfactory and other multi-sensory experience of the virtual world, so as to make people in the virtual world feel an immersive feeling. With the continuous development of social productivity and science and technology, the demand for VR technology in all walks of life is becoming increasingly strong. VR technology has made great progress and gradually become a new field of science and technology.

Internet of Things (IOT)　　Internet of Things refers to the real-time collection of any object or process that needs to be monitored, connected, and interacted with and the collection of various needed information such as sound, light, heat, electricity, mechanics, chemistry, biology, and location according to various information sensors, radio-frequency identification technology, global positioning systems, infrared sensors, laser scanners, and other various devices and technologies. It aims to achieve ubiquitous connection between things and things, things and people as well as intelligent perception identification and management of objects and processes through various types of possible network access. The Internet of Things is an information carrier based on the Internet, traditional telecommunication networks and so on which allows all ordinary physical objects that can be independently addressed to form an interconnected network.

Augmented Reality (AR)　　Augmented Reality is a technology that skillfully integrates virtual information with the real world. It makes extensive use of multimedia, three-dimensional modelling, real-time tracking and registration, intelligent interaction, sensing and other technological so as to apply computer-generated virtual information such as text, images, three-dimensional models, music, video and other virtual information simulation to the real world, where the two types of information complement each other, thus realizing "augmentation" of the real world.

Telecom Telecommunications refers to the transmission of information between different locations by using the electronic technology. Telecommunication includes different types of long-distance communication, such as radio, telegraph, television, telephone, data communication and computer network communication. Telecommunications is an important pillar of the information society. Whether in human social and economic activities or in all aspects of people's daily lives, telecommunications is an efficient and reliable means.

Read between Lines

Direction: Read the passage carefully and answer the questions below.

Question 1: Which industry that has been influenced significantly by 5G technology? (Para.1~2)

_____.

Question 2: What's the key difference between 5G and previous generations of cellular technology? (Para.3~4)

_____.

Question 3: How does 5G technology impact the real estate industry? (Para.5~6)

_____.

Question 4: What can real estate professionals do regardless of location according to 5G technology? (Para7~8)

_____.

Question 5: What are the challenges in adopting 5G technology? (Para.9~12)

_____.

Question 6: Do the advantages in adopting 5G technology exceed disadvantages in terms of the real estate industry? (Para.11)

_____.

Words and Expressions

Ex.7.1. *Write the word according to the definition. The first letter is given.*

1. h_____: a problem or difficulty that must be solved or dealt with before you can achieve something.
2. p_____: happening or existing before the event or the object that you are talking about.
3. e_____: a large area of land, usually in the country, that is owned by one person or family.
4. c_____: to work together with somebody in order to produce or achieve something.
5. a_____: a feature that makes a place pleasant, comfortable or easy to live in.

Ex.7.2. *Fill in the blanks of the appropriate word from the box below. Change the form if necessary.*

| Automated | capacity | interact | constant | potential |
| Significantly | dependable | connectivity | immersive | advanced |

1. Price is determined through the _____ of demand and supply.
2. _____ reading, by contrast, depends on being willing to risk inefficiency, goal lessness, even time-wasting.
3. The research is too unfocused to have any _____ impact.
4. Anne was intelligent and _____ of passing her exams with ease.
5. He has the _____ to become a world-class musician.
6. You don't have to pay for the tickets in _____.
7. Every _____ that you make to the network is stamped with your IP address.
8. The equipment was made on highly _____ production lines.
9. Wage rates _____ on levels of productivity.
10. Surgical techniques are _____ being refined.

Ex.7.3. *Complete each sentence below with the appropriate form of word in bracket.*

1. Why are so few companies truly _____ (innovation)?
2. These children have a huge reserve of _____ (latency) talent.
3. She undertook the task of _____ (monitor) the elections.

4. What lies behind this explosion in international _____ (commercial)?
5. The way in which we work has undergone a complete _____ (transform) in the past decade.
6. She lived through the turmoil of the French _____ (revolutionize).
7. The key doesn't fit the _____ (unlock).
8. When she saw the house, she had a feeling of _____ (familiar).
9. Administrative staff may be deskilled through increased automation and_____ (efficient).
10. Many people are unaware of just how much food and drink they _____ (consumption).

Ex.7.4. *Translate the following Chinese sentences into English.*

1. 随着 5G 技术的普及，蜂窝网络的覆盖范围不断扩大，为移动设备和物联网设备提供了更稳定、更快捷的数据连接。(cellular)

   _____.

2. 是否建立这个专用网络的计划，取决于能否以负担得起的价格建立可靠的网络基础设施。(infrastructure)

   _____.

3. 社会科学家和研究人员表示，短信正在改变我们的互动方式。(interact)

   _____.

4. 计算机在处理信息方面效率极高。(efficient)

   _____.

5. 大多数情况下，涉及浏览器的部分我们都有合作。(collaborate)

   _____.

Theme Writing

Directions: You are required to write a passage of about 80 words to summarize the application of 5G technology according to the following passage.

Passage:

China has been expanding its 5G infrastructure at a rapid pace, utilizing next-generation wireless technology to facilitate the digitalization of different sectors of the country.

The rapidly expanding telecom network, featuring high speed, high reliability and low latency, has enabled futuristic scenarios in sectors like manufacturing, medical services and farming.

Smart factories

China's leading automotive lithium-ion battery maker Contemporary Amperex Technology Co., Ltd. (CATL) is digitalizing its battery plants. The centrally controlled production, ultrahigh speed visual quality inspection and augmented-reality expertise system at the CATL production plant are all 5G enabled. 5G technology allows CATL engineers to monitor the entire workshop in real time and quickly handle technical issues remotely as if they were on site. Cell coating is one of the most delicate processes in battery production since materials must be evenly spread on micrometer-thick copper or aluminum foils. In the workshop, employees use tablet computers to control the extrusion heads of coating machines at full pace. A two-story-high coating machine, measuring 100 meters in length, can handle a 100-meter-long foil per minute, producing one cell in 1.7 seconds. Its defect rate is recorded as merely one in a billion, thanks to fast and stable 5G transmission. The factory also has more than 200 5G-enabled automated vehicles shuttling through its assembly lines to transport batteries and other materials. CATL plans to implement 5G technology in its 10 global production bases, said Chen Ling, the company's chief information officer. The Jiangnan Shipyard under the China State Shipbuilding Co., Ltd. in Shanghai is also embracing 5G. The state-owned shipbuilder is working with scientists from Shanghai Jiao Tong University to develop a 5G-enabled smart system to shorten the working process in shipbuilding from 14 hours to about 2 hours. Intelligent production facilities in China such as CATL and Jiangnan have continued to proliferate, making high-tech manufacturing a trend.

Remote surgeries

In 2019, a surgeon at Shanghai-based Huashan Hospital performed an endoscopic surgery to remove a pituitary tumor in a patient's brain. About 20 kilometers away, nearly 60 interns closely watched the operation, which was livestreamed on a 4K high-definition screen. Thanks to 5G wireless devices developed by Chinese telecom giant Huawei, such cutting-edge training was possible. It marked the launch of Shanghai's first 5G smart medical pilot base. Surgical procedures in China are evolving due to the 5G network's low latency advantage. On Jan. 17th this year, surgeons at Huashan Hospital made an attempt to excise a basicranial tumor, bolstered by 5G mixed reality (MR) technology developed by a Shanghai-based startup. Wearing VR headsets, they cut out target tissues precisely through a five-centimeter incision.

The brain's enhanced 3D holographic images were projected in real-time upon the patient's scalp to indicate exact locations inside the brain. Huashan's 5G+MR surgery was listed among the Top 10 Cases at this year's World 5G Convention held in Harbin, Heilongjiang province, in early August.

5G technology also supports remote-controlled robotic operations. Medical workers from two hospitals in the eastern Chinese city of Nanjing and the westernmost prefecture of Kezilesu Kirgiz in the country's Xinjiang Uygur autonomous region collaborated to conduct surgery for stone removal through a 5G-enabled robot. Experienced surgeons in Nanjing, Jiangsu province, used a long-distance ultrasound machine to give instructions on the precise place of the puncture, while their counterparts 5,000 km away operated accordingly to remove the stone. During the operation, the average two-way 5G network latency was reduced to only 135 milliseconds, meaning the surgeons could communicate without any noticeable time delay. As of Aug. 1st, the Kezilesu Kirgiz hospital had completed 22 stone removal robotic operations, making high-quality medical services more accessible to patients living in the country's remote areas.

Digital farming

Agriculture, commonly seen as a natural resource, has also been integrated with 5G technology. At a rice paddy field in Northeast China's Heilongjiang province, a probe equipped with a 5G-enabled AI camera is used for leaf age diagnosis, disease identification, pest control and prevention of plant diseases. Along lush crop rows, water-level sensors are installed to measure soil moisture and enable automatic irrigation. "The 5G digital farm has improved tillage, making agriculture more manageable," said Meng Qi, a contractor of the smart farm. "5G agricultural machinery, the internet of things, artificial intelligence and other digital technologies are becoming the new tools for rural development," said Li Daoliang, director of the International College Beijing under China Agricultural University.

第二章 拓展阅读

Reading 1

China Builds World's Largest 5G Network with 475 Million Users

1 China has built nearly 1.97 million 5G base stations and reached 475 million 5G mobile users by the end of this July, said the country's Ministry of Industry and Information Technology

(MIIT), constituting the world's largest 5G network.

2 The 5G base stations currently in operation in China account for over 60 percent of the world's total three years after the issuance of 5G commercial licenses, and the country's 5G users make up more than 70 percent of global 5G users. The construction of 600,000 5G base stations will be completed this year, the MIIT said.

3 China's 5G commercial license has been officially issued for three years, with record-breaking base stations scale and constant breakthroughs in key technologies. The 5G development in China not only promotes leapfrog growth of the information and communication industry, but also injects a strong impetus to the digitalization of the economy and society.

4 "The 5G connection was poor on the ship when I came here last year, but this time I'm always connected," said a man surnamed Yang taking his family to thc Putuo Mountain, one of the four famous Buddhist mountains in China located in east China's Zhejiang province. On the ship to the mountain, Yang could even watch sports games in high definition on his mobile phone.

5 The improved network experience was attributed to a 5G base station built on an uninhabited island in Zhoushan, Zhejiang province. Following coordinated plans and taking a moderately proactive approach to advancing infrastructure investment, Chinese telecommunication carriers have constantly expanded the number of 5G base stations.

6 According to statistics, by 2021, all cities above the prefecture level across China have been covered by the 5G network, while the 5G signal has been available in over 98 percent of county seats and 80 percent of townships.

7 China Unicom and China Telecom, two major carriers in China, jointly built the world's first co-shared 5G SA network, which has now achieved coverage in urban areas, counties and key townships. China Mobile, another major carrier in the country, has built over 200,000 base stations in rural areas in cooperation with China Broadnet, a new telecommunication carrier.

8 The co-construction and sharing have not only speeded up the building of the 5G network, but also tremendously saved capital and resources, contributing to the green and low-carbon operation of enterprises.

9 It is reported that the co-construction of 4G and 5G networks by China Unicom and China Telecom alone has saved over 210 billion yuan ($30.65 billion) of investment in the past two years. It is expected save over 20 billion yuan in operation costs, 10 billion kilowatt-hours of electricity consumption and 6 million tons of carbon emission annually.

10 Apart from the accelerating network construction and sharing, China has made breakthroughs in technological innovation in key 5G technologies. It has constantly narrowed the gap with global industry leaders in terms of 5G chips, mobile operating systems and other core technologies. For instance, China Mobile has led 156 international 5G international standard projects and applied for over 3,600 5G patents, which makes it one of the leading carriers around

the globe.

11　While making constant progress in 5G technologies, China has also strengthened its 5G industry. In 2021, the country shipped 266 million 5G phones, up 63.5 percent year on year. As of April this year, there were 1,334 models of 5G devices in the world, including 677 mobile devices. Over 80 percent, or 558 of these mobile devices came from China.

12　The 5G technology, with large bandwidth, low latency, wide connectivity and other features, is being merged with technologies such as high-definition videos, augmented reality, virtual reality and artificial intelligence, to create massive application scenarios in the audio and video sectors.

13　An MIIT official said China will further encourage innovation and solidify the industrial foundation to promote application innovation, technological breakthroughs and ecological construction in the field of 5G.

People's Daily 2022-08-26

Comprehension Check for Reading 1

Direction: Read the passage carefully and answer the questions below.

1. What is the main idea of this passage?

　　A) 5G network has more advantages in comparison to 4G network.

　　B) The co-construction and sharing have speeded up the building of the 5G network.

　　C) China has constantly made significant progress in terms of 5G network.

　　D) 5G base stations now in operation account for over 60 percent of the world's total.

2. Which of the following statement is TRUE according the passage?

　　A) China has owned the second largest 5G network.

　　B) China Unicom built the world's first co-shared 5G SA network.

　　C) China Mobile has led 156 international 5G international standard projects and applied for over 3,000 5G patents.

　　D) The 5G signal has been available in over 98 percent of county seats and 80 percent of townships by 2021.

3. Which of the following is not included in the features of 5G technology?

　　A) Low price.　　B) Large bandwidth.　　C) Wide connectivity.　　D) Low latency.

4. Which company does China Broadnet has cooperated with to gain the result of building over

200,000 base stations in rural areas?

   A) China Unicom.

   B) China Mobile.

   C) China Telecom.

   D) Not mentioned.

5. What will China do to promote application innovation, technological breakthroughs and ecological construction in the field of 5G?

   A) Completing the construction of 600,000 5G base stations.

   B) Encouraging innovation and solidifying the industrial foundation.

   C) Accelerating network construction and sharing.

   D) Following coordinated plans and taking a moderately proactive approach to advancing infrastructure investment.

Reading 2

Will the Cloud Business Eat the 5G Telecoms Industry?

As AT&T and Verizon launch 5G this month, two huge industries collide

1   Smartphones able to take advantage of zippy fifth-generation (5G) mobile telephony have graced American pockets since 2019. Samsung launched its first 5G-enabled device in April that year. Apple followed suit in late 2020 with its long-awaited 5G iPhone. Until now, however, actual 5G coverage in America has been limited. The country's three biggest carriers, AT&T, Verizon and T-Mobile, have offered 5G connectivity but in practice this differed little from the earlier 4G. AT&T and Verizon had to delay their large-scale roll-outs of something closer to the hype in December after the Federal Aviation Administration aired concerns that their 5G radio spectrum interferes with avionics on some ageing aircraft. On January 3rd both firms, which insist that the technology is safe (and can be turned off around airports, just in case), said they would again postpone switching on their 5G networks by two weeks.

2   Yet it is the imminent arrival of another player in the 5G contest that is the talk of the industry. In the next few months Dish Networks, a company best known for its satellite-television service, is expected to launch America's fourth big carrier. The firm's promise to inject more competition into a concentrated and ossified sector was what helped persuade regulators to approve a merger between T-Mobile and Sprint, a smaller incumbent, in 2020.

3   More important, Dish's network is to be the first in America that would live almost entirely in a computing cloud. Except for antennas and cables, it is mostly a cluster of code that runs on Amazon Web Services (AWS), the e-commerce giant's cloud-computing arm. As such, the roll-out is a test of the extent to which computing clouds will "eat" the telecoms industry, as software has eaten everything from taxis to Tinseltown. If the launch is a success and other carriers follow suit, it could reconfigure not just America's wireless industry but the global mobile-telecoms market with annual revenues of around $1trn, according to Dell'Oro Group, a research firm. And it would entangle telecoms intimately with the cloud business, whose revenues could be half as large this year and are growing at double digits.

Dish best served cold

4   Dish's network is the culmination of a process that started in the early 1980s. Back then antitrust regulators allowed AT&T, the world's largest network operator, and IBM, its biggest computer firm, to enter each others' markets. AT&T started selling personal computers and IBM bought ROLM, which sold telecoms equipment. Pundits predicted an epic battle between the two giants — and a rapid convergence of the telecoms and computer industries into one.

5   Neither the battle nor the convergence materialized. Forty years ago the two markets proved too distinct and the technology was not up to snuff. Now things look different. Computing clouds such as AWS and Microsoft's Azure are maturing fast, and finally becoming able to deal with the demanding task of powering a mobile network. The latest iteration of mobile technology, 5G, was conceived from the start not as a collection of switches and other hardware, but as a set of services that can be turned into software, or "virtualised". And the telecoms industry is becoming less proprietary, embracing "open radio access network" (o-ran) standards that make it possible to virtualise ever more functions previously performed by hardware. As a result, networks can turn into platforms for software add-ons, just as mobiles turned into smartphones which could run apps.

6   All this will be on full display in Dish's network. Instead of bulky base stations used in conventional mobile networks, its technology is housed in slender boxes attached to antenna posts. These are connected directly to the AWS cloud, which hosts the virtual part of the network, including all of Dish's other software (for example that used to manage subscribers and billing). The only thing Dish is buying from established makers of telecoms gear is software, says Marc Rouanne, its chief network officer (who used to work for one such vendor, Finland's Nokia).

7   As a result, Dish's network will be cheaper to set up and to run. It will also be fully automated, down to the virtual "labs" where new services are tested. This should allow the company quickly to spin up special-purpose networks, for instance connecting equipment in mine shafts, or enabling drones to talk to each other and their controllers. Dish also wants to use

artificial intelligence to optimise the use of radio spectrum, including by training algorithms which are able to adapt parts of the network to specific conditions such as a storm or a mass concert.

8   Although Dish is pushing this "cloudification" furthest, other carriers around the world are not far behind. In June AT&T, still America's largest mobile operator, sold the technology that powers the core of its 5G network to Microsoft, which will run it for AT&T on its Azure cloud. Reliance Jio, India's technology titan, has ambitious plans to build a cloud-based 5G network.

9   These developments are also bringing the big cloud providers into the telecoms world. Last year Microsoft bought Affirmed Networks and Metaswitch, the main software suppliers for the core of AT&T's 5G network. They now form a new business unit called "Azure for Operators". Google has a similar effort and recently forged a partnership with Telenor, a Norwegian telecoms company. In November AWS announced a new offering that lets customers quickly set up private 5G networks on their premises.

10   Newcomers are also elbowing their way into the business. Rakuten, a Japanese online giant, has already built a Dish-like network at home. Rather than outsourcing its cloud operation to big tech, Rakuten has built its own, and launched a subsidiary, called Rakuten Symphony, to offer the system to other operators. It is helping 1&1, a German web-hosting company, to build a network. "We don't want to be a telco cloud, but enable operators to make their own," explains Tareq Amin, who heads Rakuten Symphony.

11   Existing mobile networks will not be replaced overnight. Rakuten's faced delays and Dish's was originally scheduled for launch late last year. Some technical barriers remain. Despite being seen as a welcome alternative to gear from Huawei, a controversial Chinese giant, especially in Europe, gear based on o-ran specifications is not mature. Its European adopters have therefore yet to install it in the most vital parts of their networks. "It's in an extended beta test," sums up Dean Bubley of Disruptive Analysis, a consultancy.

12   Another question is whether the cloud can completely gobble up telecoms networks, notes Stéphane Téral of LightCounting, another consultancy. Controlling a 5G base station is hugely complex and involves keeping tabs on hundreds of para meters. The more flexible a carrier wants to be, the more complicated things get. At least for some time, the necessary control software may need to run on specialised gear near the antenna rather than on generalist servers in faraway data centres.

13   Then there are the political and financial barriers. European governments fret that America's spooks will have even more access to their country's networks if these run in American clouds (Europe has none of its own and is understandably even warier of Chinese ones). Carriers, in Europe and elsewhere, fear losing business to the tech giants like Amazon, Google or Microsoft, which have already skimmed most of the value generated by 4G mobile technology. "If all this is not financially interesting for [telecoms firms], they will try something

else," says Michael Trabbia, chief technology officer of Orange, a French mobile operator.

14 However this plays out, the telecoms business will look very different a few years from now. The contest for control of the telecoms cloud, and particularly its "edge" (tech speak for what remains of the base station) will only heat up. Whoever is in charge of these digital gates will have the fastest access to consumers and their data, the main currency in a world of new wireless services, from self-driving cars to virtual-reality metaverses.

15 The cloud businesses have the technological edge for now, and will try to eat as much of wireless networks as possible. The operators have relationships with customers, know how to manage networks and own the radio spectrum. Eventually, cloud providers and network operators will probably come to some kind of agreement. In the new world of mobile telecoms, neither can do without the other.

16 Clarification (January 7th 2022): An earlier version of this article suggested that only T-mobile has offered some 5G connectivity. In fact all three of America's big carriers do, though their existing networks offer only limited 5G functionality.

Comprehension Check for Reading 2

Direction: Skim the article and answer the following questions.

1. Why did AT&T and Verizon have to delay their large-scale 5G network service?

_____

2. What happened to the industries of telecoms and computer in 1980s?

_____

3. What are the advantages of the Dish Networks in the 5G competition?

_____

4. Why will the existing mobile networks will not be replaced overnight?

_____

5. According to the passage, will the cloud business eat the 5G telecoms industry?

_____

Unit Eight

Virtual Reality Technology

——— 第八单元 ———

虚拟现实技术

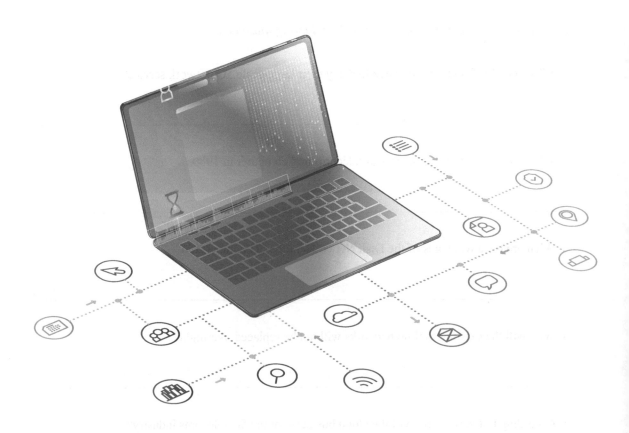

第一章　主题阅读

> **Text**

VR Continues to Make People Sick - and Women More So than Men

1　The whole history of fiction shows that alternative realities are an attractive and profitable idea. So back in the 1990s, when electronics had arrived at a point where people could build headsets that blocked off actual reality and replaced it with a virtual version created inside a computer, it looked as if something world-changing might have arrived. Games companies were particularly excited, and Nintendo, Sega and Virtuality duly piled in.

2　The world, however, stubbornly refused to be changed. It might have put up with the low-resolution images, the choppy scene transitions and the poor controls, for these would surely have got better. It might also have put up with the price (the headsets in question could cost up to $70,000), for that would surely have come down. It could not, though, accommodate the dizziness, nausea, eye strain, vomiting, headaches, sweating and disorientation that many of the technology's users (more than 60%, according to one study) complained of a set of symptoms that, collectively, have come to be called "cyber-sickness". Though not fatal to people, cyber-sickness certainly helped damage the industry, which was more or less vanished.

3　Two decades later, however, virtual reality (VR) returned from the dead, with better images, smoother transitions and more precise controls. There were also applications beyond games. The upgraded technology has found use in social media, interior design, job training and even pain management. Moreover, a new set of companies, Oculus, HTC and Sony, have come up with products that do not require a second mortgage to afford.

4　Despite these improvements, though, VR has not lived up to expectations. It will have done respectably, with sales in 2022 of $3.1bn, according to Omdia Analysis, a market research firm. But that is only 1.7% of the global market for games. Many people thought that this time around VR would become a blockbuster technology. It has not happened. Part of the reason is that cyber-sickness has not gone away. One study suggests between 25% and 40% of users still experience it.

5　Dealing with this is difficult, not least because there is an argument about what triggers it in the first place. Two theories dominate. One is that users experience sensory conflict — a mismatch between what they see and what their other senses and their real-world knowledge tell them they should be experiencing. The other is that the underlying cause is individuals' inability to control their bodies and maintain proper posture when moving around in virtual environments. To complicate matters, both hypotheses could be true.

6   Sensory conflict there certainly is. For example, when users move their heads they expect what they see to change immediately in response. But time-lags and poor graphics mean their visual input often fails to meet the brain's expectations. Dealing with this means increasing the "frame rate" at which the virtual world is presented to a user, improving the resolution of the images and reducing the latency of response to a user's movements. All of these require clever processing by the computer responsible for creating the illusion.

7   Improvements in tracking what a user are doing also help. "Room scale" VR systems let people move around in the real world while perceiving similar movement in the virtual one. Following a user's movement can be done in one of two ways. Outside-in tracking relies on external cameras observing beacons of various sorts scattered around a user's body. Inside-out tracking is the opposite: the beacons are scattered around the room and detectors on a user's body employ them as reference points.

8   On top of all this, there is the design of the lenses that sit inside a headset in front of a user's eyes to adjust optically for the fact that what is actually a nearby image is supposed to be some distance away. Since the shape of these lenses is fixed and the amount of adjustment required varies with what is being looked at, distortion is inevitable. But distortions are particularly noticeable when users move their eyes, says Paul MacNeilage of the University of Nevada, Reno. Some headsets therefore now track a user's gaze and move the lenses within the headset in response.

9   Make the input too credible, though, and you run into a different problem-the contrast between what a user's eyes are seeing and what the motion-sensors in his inner ear are detecting. To deal with that, some designers program in a "virtual nose", just visible to the user and to serve as a point of reference.

10   These tactics help. But they do not get rid of cyber-sickness entirely. That is where the second hypothesis, unstable posture, comes in. And it is one that has the virtue of offering an explanation of a mystery about the condition — why women are more likely to be affected than men.

11   Thomas Stoffregen of the University of Minnesota, who has studied the matter and found women four times as susceptible as men, cites the example of driving a car to explain the unstable-posture hypothesis. When turning the steering wheel, he observes, drivers need to keep their heads oriented to the road. They need to stabilize their bodies, particularly when the car is changing direction and pushing the body in different ways. "When you spend a lot of time in cars, you get used to doing that," he says. "It's a skill." But in virtual environments,

where there are no forces to act as signals, people have not learned to adjust their bodies properly. They lean when the virtual car turns, but in fact they are leaning away from stability. He finds this particularly affects women, who have lower centres of gravity than men. That may cause them to sway more. And increased swaying, he has found, correlates with higher rates of cyber-sickness.

12 It is a neat idea. But Bas Rokers of the University of Wisconsin-Madison believes there is a simpler explanation for women's experience of cyber-sickness, which is that headsets are not designed for them. For VR to work properly, sets need to be adjusted to the distance between the pupils of a user's eyes. In one popular brand, however, Dr. Rokers found that 90% of women have an interpupillary distance less than the default headset setting, and 27% of women's eyes do not fit the headset at all.

13 If Dr. Rokers is correct, a big part of the problem of cyber-sickness might be dealt with by a small change to helmet design. If women's rates of the complaint could be reduced to the level experienced by men, then a lot more people could enjoy VR rather than enduring it. And then, perhaps, it really might achieve its potential.

The Economist 2022-03

Vocabulary

accommodate *v.* 为……提供住宿；容纳；考虑到，顾及；顺应，适应
adjust *v.* 调整，调节；适应，习惯
alternative *n.* 可供选择的事物，替代物；*adj.* 可替代的，备选的
application *n.* 正式申请，书面申请；申请书；敷用；应用，实施
beacon *n.* 灯塔；(无线电)信标
block off 阻塞(堵塞某个通道或道路，使其无法通过)
blockbuster *n.* 一鸣惊人的事物，(尤指)非常成功的书(或电影)
cite *v.* 引用，援引；引证，引以为例
choppy *adj.* 波涛汹涌的
complicate *v.* 使复杂化，使难以理解；引起并发症；使卷入，使陷入
contrast *n.* 差异；*v.* 形成对比

correlate *v.* 相互关联；*n.* 相互关联的事物
cyber-sickness *n.* 晕动症
default *n.* 不履行；默认；*v.* 违约；*adj.* 默认的
detector *n.* 探测器，检测器
distortion *n.* 歪曲，曲解；变形，失真
dizziness *n.* 头晕；头昏眼花
dominate *v.* 统治，支配；在……中占首要地位；俯视，高耸于；占绝对优势
endure *v.* 持续存在，持久；忍受维度
expectation *n.* 期待，预期；期望，指望
fatal *adj.* 致命的；导致失败的，灾难性的
fiction *n.* 小说；虚构的事，谎言；杜撰，编造
graphic *adj.* 详细的，生动的；绘画的；*n.* 图表，图形；绘画，图形设计
gravity *n.* 重力，地心引力；严重性；严肃，庄严；重(量)

headset *n.* 耳机；头戴式受话器
helmet *n.* 头盔，安全帽
hypothesis *n.* 假说，假设；猜想，猜测；前提 *pl.* hypotheses
illusion *n.* 错觉，幻觉；幻想，错误的观念
interior *adj.* 内部的，里面的；国内的，内政的；*n.* 内部，里面；内陆
interpupillary *adj.* 瞳孔间的
lean *v.* (身体) 倾斜；倾向于做；倚靠；*n.* 倾斜，歪曲；瘦肉
mortgage *n.* 按揭，抵押贷款；抵押贷款额 *v.* 抵押；
mystery *n.* 难以理解 (或解释) 的事物；*adj.* 神秘的，身份不明的
nausea *n.* 恶心，呕吐感
optically *adv.* 光学地；眼睛地
oriented *adj.* 以……为方向的，重视……的；*v.* 朝向，面对，使适合；
perceiving *n.* 感知；熟思型
potential *adj.* 潜在的，*n.* (事物的) 潜力，可能性；(人的) 潜能，潜力
profitable *adj.* 盈利的，有利可图的；有益的，有用的
respectably *adv.* 相当好地；
scatter *v.* 撒，播撒；(使) 散开
sensory *adj.* 感觉的，感官的
stabilize *v.* (使) 稳定，稳固
strain *n.* 焦虑，紧张；*v.* 拉伤，扭伤
stubbornly *adv.* (人) 固执地
susceptible *adj.* 易得病的
sway *v.* 摇摆，摇晃，摆动
symptom *n.* (医) 症状
tactic *n.* 策略，手法；战术
time-lag *n.* 时间间隔
track *n.* 小道；踪迹；*v.* 追查，追踪
transition *n.* 过渡；*v.* 转变
trigger *v.* 引发，激发；*n.* 诱因
underlying *adj.* 根本的，潜在的
visible *adj.* 看得见的；明显的
vanish *v.* 突然不见，消失

Cultural Notes

Nintendo Nintendo is a Japanese company mainly engaged in the development of electronic game software and hardware, one of the three giants in the electronic game industry, and the pioneer of the modern electronic game industry. In 1995, Nintendo produced a game console called Virtual Boy which is the first home VR device in the world.

Sega Sega is a Japanese electronic game company that has simultaneously produced home game console hardware and its corresponding game software, business game console hardware and its corresponding game software, and computer game software. In 1991, Sega announced the launch of VR glasses. This pair of glasses can be used for arcade games.

Virtuality Virtuality Group, a British company, launched "Virtuality", which was the first VR gaming console in 1991. Virtuality has shocked the entire industry with its brand new immersion, and it is also the first large-scale production in the history of virtual reality entertainment. This machine can support network and multiplayer games and is equipped with a series of hardware devices, such as virtual reality glasses, graphics rendering systems, 3D trackers, and wearable devices similar to exoskeletons.

Oculus   Oculus VR is a virtual reality technology company based in Irvine, California, founded in 2012. It was acquired by Facebook in 2014 and its first product was Oculus Rift. In 2019, Facebook launched the Oculus Quest all-in-one machine. In 2021, Facebook was renamed Meta, which is the prefix for the term Metaverse.

HTC   HTC International Electronics Co., Ltd., founded in 1997, is a mobile phone and tablet computer manufacturer located in Taiwan, China. It is the world's largest manufacturer of Windows Mobile smartphones, as well as the world's largest smartphone foundry and manufacturer. In 2015, HTC Vive glasses developed by HTC and Valve made their debut at the World Mobile Conference.

Sony   Sony is a globally renowned large comprehensive multinational enterprise group in Japan. It is a global manufacturer of audiovisual, electronic games, and communication products. In 2003, Sony has released EyeToy for PlayStation 2, a digital camera used for gesture recognition. This device allows players to interact with the game through body posture, color, and even sound, and it also has a built-in microphone. Although EyeToy did not achieve commercial success, it allowed Sony to enter the virtual reality market. In 2014, Sony announced the launch of a virtual reality device called Project Morpheus for PlayStation 4.

Second Mortgage   Second Mortgage is translated as "second priority mortgage". Multiple common law or equity mortgages can be established on the same immovable property, with the second priority mortgage being second only to the first priority mortgage. The position of the second ranked mortgagee is second only to that of the first ranked mortgagee. The first priority mortgagee has the right to sell the mortgaged property without the consent of the second priority mortgagee, and all mortgage rights of the mortgaged property that have been sold shall be extinguished. The proceeds from selling the collateral must first be used to pay off the bonds of the first priority mortgagor, and if there is any surplus, the bonds of the second priority mortgagor must be used to pay off.

Read between Lines

Direction: Read the passage carefully and answer the questions below.

Question 1: What are common symptoms of "cyber-sickness"? (Para. 2)

_____.

Question 2: Why VR hasn't lived up to people's expectations? (Para.4)

_____.

Question 3: What causes "cyber-sickness"? (Para.5)

_____.

Question 4: How can we solve a user's sensory conflict? (Para.6～7)

_____.

Question 5: Why VR headsets are prone to distortion? (Para.8)

_____.

Question 6: Why are women more susceptible to cyber-sickness than men? (Para.11～12)

_____.

Words and Expressions

Ex8.1. *Write the word according to the definition. The first letter is given.*

1. d_____: the changing of the appearance or sound of something in a way that makes it seem strange or unclear.
2. a_____: a thing that you can choose to do or have out of two or more possibilities.
3. p_____: that makes or is likely to make money.
4. b_____: something very successful, especially a very successful book or film/movie.
5. n_____: the feeling that you have when you want to vomit, for example because you are sick or are disgusted by sth.
6. s_____: to move or to make people or animals move very quickly in different directions.

Ex.8.2. *Fill in the blanks of the appropriate word from the box below. Change the form if necessary.*

accommodate	lean	correlate	endure	block off
perceiving	default	vanish	adjust	susceptible

1. The study of an academic discipline alters the way we _____ the world.
2. I felt I had _____ to the idea of being a mother very well.
3. Snow has _____ traffic on the motor way.
4. She walked slowly, _____ on her son's arm.
5. After the earthquake, the first thing the local government did was to provide _____ for the homeless families.
6. Such stereotypes about spoiled only children have _____ for more than a century.
7. Mortgage _____ have risen in the last year.
8. It's on the stage in the middle of the journey that people feel youth _____ and their prospects narrowing.
9. There is a direct _____ between exposure to sun and skin cancer.
10. Obesity raises _____ to cancer, and Britain is the sixth most obese country on Earth.

Ex.8.3. *Complete each sentence below with the appropriate form of word in bracket.*

1. His _____ (apply) for membership of the organization was rejected.
2. I do not wish to _____ (complication) the task more than is necessary.
3. Good sound science depends on _____ (hypothesis), experiments and reasoned methodologies.
4. Dress formally and _____ (respectable) in upscale restaurants, and ensure that your table manners are impeccable.
5. Women have superior _____ (sense) abilities compared to men.
6. If your smoke _____ (detect) is working properly, the red light should be on.
7. Vocational-type classes, such as computer science or journalism, are often more research-_____ (orient) and lend themselves to take-home testing.
8. He kept telling us about his operation, in the most _____ (graph) details.
9. He _____ (stubborn) refused to tell her how he had come to be in such a state.
10. The rain had stopped and a star or two was _____ (visibility) over the mountains.

Ex.8.4 *Translate the following Chinese sentences into English.*
1. 东西方文化之间存在着明显的差异。(contrast)

_____.

2. 在当前的人工智能领域，深度学习技术占据了主导地位，推动了诸多创新应用的发展。(dominate)

_____.

3. 这项新技术在计算机安全领域具有巨大的潜力，能够有效防御网络攻击。(potential)

_____.

4. 业界对这款新开发的人工智能软件寄予厚望，期待它能大幅提升工作效率。(expectation)

_____.

5. 这款新型计算机的内部设计采用了最新的散热技术，确保了长时间运行的稳定性。(interior)

_____.

Theme Writing

Directions: You are required to write a passage of about 120 words based on your understanding of the short passage below. Your writing should respond to the following questions:
1) What do you know about VR?
2) How do you understand the underlined sentence, "Your body is not in the VR world, but your spirit is"?

Passage:

VR is a simulated experience that can be similar to or completely different from the real world. It's a technology that transports people to another world that may be real or fictional by its creators. I would like to call the world of VR a constructed world. When you put on the VR device, you can immediately enter this constructed world.

The breakthrough significance of VR technology is that it allows people to "immerse themselves", which is the main different from any other technology. Your body is not in the VR world, but your spirit is. For example, when you put on your VR device and watch the VR interactive film "Notes on Blindness - Into Darkness" based on the audio diary of the scholar John Hull, you can feel the importance of the sense of hearing in the world of the blind and also the loneliness and fear of blind people diving in the dark. As for, you can also feel the same feeling with VR glasses at the moment.

第二章 拓展阅读

Reading 1

Tech Giants Bank on VR for Metaverse Opportunities

1　HTC Corp will step up popularizing the use of virtual reality technology in both the consumer market and the enterprise market as part of its broader push to better tap into the metaverse, a senior executive said.

2　As a hot tech buzzword, metaverse (元宇宙) refers to a shared virtual environment or digital space created by technologies including virtual reality and augmented reality.

3　Charles Huang, corporate vice-president of HTC, said there is surging demand for VR devices in a wide range of segments, such as remote conferencing, training, education, healthcare, design and exhibition, with the COVID-19 pandemic bolstering the application of VR equipment.

4　According to Huang, HTC is banking on the business-to-business segment and attaches great importance to cooperation with industry partners to develop new applications.

5　"Metaverse is not a slogan or a very distant future," Huang said, adding that HTC is working hard to strengthen technological exchanges and information sharing in the metaverse and striving to build an interconnected, mutually beneficial and compatible metaverse ecosystem.

6　HTC has unveiled an open metaverse platform called Viverse, which provides seamless experiences that are reachable on any device anywhere and is enabled by VR, AR, high-speed connectivity, AI and block chain technologies.

7　Data from market consultancy International Data Corp showed that the investment scale of the global AR and VR market was close to $14.67 billion in 2021 and is expected to increase to $74.73 billion in 2026, with a compound annual growth rate of 38.5 percent.

8　IDC said the IT-related expenditure in China's AR and VR market reached about $2.13 billion in 2021 and will increase to $13.08 billion in 2026, making it the second-largest market in the world.

9　Pedro Palandrani, a technology analyst at research company Global X ETFs, said early versions of metaverse now exist, offering investors a glimpse of its enormous potential. However, a successful metaverse is expected to feature a decentralized open platform accessed by VR headsets and powered by block chain technology.

10 "A truly immersive metaverse experience, for instance, engages all the senses — sight, sound, touch, smell and taste. Today, VR mostly involves sound and images," Palandrani said.

11 To accelerate the development of metaverse, other Chinese tech companies are also moving fast. Huawei and Alibaba, for instance, are among the first group of companies, along with Meta, formerly known as Facebook, and Microsoft from the United States, to form a standards group that aims to accelerate the development of metaverse.

12 Participants in the Metaverse Standards Forum include many of the biggest companies working in the sector, from chipmakers to gaming companies, as well as established standard-setting bodies like the World Wide Web Consortium, according to the group.

13 "Industry leaders have stated that the potential of metaverse will be best realized if it is built on a foundation of open standards," the group said. "Building an open and inclusive metaverse at a pervasive scale will demand a constellation of open interoperability standards."

14 The move signified that companies are racing to build the concept of a metaverse and want to make their nascent digital worlds compatible with each other, said Pan Helin, co-director of the Digital Economy and Financial Innovation Research Center at Zhejiang University's International Business School.

15 That could make it easier for developers to build the same content for different metaverse platforms or for users to export data from one service to another, which will help build a thriving ecosystem in the future, Pan said.

16 Seven of China's big technology companies, including Huawei, Tencent, Baidu, Oppo and Alibaba, are among the top companies that filed the most virtual reality and augmented reality patent applications globally in the last two years, said Singapore-based research and development analytics provider PatSnap.

17 Bloomberg Intelligence forecasts that metaverse revenue globally could reach nearly $800 billion in 2024.

China Daily 2022-12-19

Comprehension Check for Reading 1

Direction: Skim the article and answer the following questions.

1. Which of the following is NOT true about metaverse?

　　A) It is enabled by VR, AR, AI and other technologies.

　　B) It is a popular technological term.

　　C) It is a technical construction of the universe.

　　D) It is a virtual world that can interact with the real world.

2. What can Viverse unveiled by HTC offer users?

    A) Convenient devices.   B) Multiple technologies.

    C) Open platform.   D) Gapless experiences.

3. What senses are involved in virtual reality?

    A) Sight and sound.   B) Sound and touch.

    C) Images and smell.   D) Sound and taste.

4. How can the potential of the metaverse be best realized?

    A) The participation of Chinese technology companies.

    B) The base on open interoperability standards.

    C) The establishment of standards groups.

    D) The integration with the digital world.

5. What is a thriving metaverse ecosystem like?

    A) Developers can fully understand the needs of users.

    B) Developers can build various contents for different metaverse platforms.

    C) Users can derive data from one service to another.

    D) Users can apply for patents for virtual reality and augmented reality.

Reading 2

Metaverse No More? ByteDance and Tencent Scale back VR Ambitions

    1   Two years after ByteDance acquired Chinese VR (virtual reality) headset manufacturer Pico for a whopping $1.3 billion, the tech giant is now reportedly downsizing and restructuring its VR division. The move comes as enthusiasm for the metaverse has waned, and emerging technologies like ChatGPT have surged to the forefront, offering transformative potential for the tech industry.

    2   ByteDance's overhaul follows in the wake of Tencent's strategic in its XR (extended reality) development earlier this year, moving away from in-house hardware production. Tencent is reportedly partnering with Meta Platforms to serve as the exclusive distributor of new, more affordable VR headsets in China with sales expected to begin in late 2024. So, two of the highest-profile players in China's VR market are scaling back.

    3   Meanwhile, Meta's Quest 3 VR headset also failed to meet expectations in the U.S. The latest survey by analyst Ming-Chi Kuo indicates that Meta has reduced its Q4 shipment volume

for Quest 3 by approximately 5%～10% this year, after market demand fell short of expectations. Apple, which only held the stunning reveal of its Vision Pro headsets in June, is reportedly cutting production forecasts of the device, and delaying the launch plans for a lower-priced, mass-market Vision product.

4　The VR industry appears to be at another pivotal juncture. Despite significant investments, the anticipated "singularity" moment of widespread adoption remains elusive, and the industry is still seeking the validation that has yet to materialize with consumers. ByteDance, in response to the uncertainty in the sector, has chosen a strategic path of conservation. The company aims to sustain its research endeavors at a minimal cost, while awaiting a potential resurgence in the VR market at a later date. The Beijing-based company is seeking to balance the need for innovation against the realities of a challenging market.

ByteDance's $3 Billion Bet on Pico

5　ByteDance is the most committed Chinese investor in the virtual reality (VR) sector. In a strategic move that underscored its obligation to VR, ByteDance acquired Pico in August 2021 for a staggering 9 billion yuan ($1.3 billion). This acquisition price significantly exceeded the market consensus, which valued Pico at around 2-3 billion yuan at that time. Following the acquisition, ByteDance kept up its pace of investment. Reports suggest that the company injected an additional 10 billion yuan in the subsequent year, aiming to boost the sales of Pico headsets. In total, ByteDance's expenditure on its VR ambitions is estimated to be around 20 billion yuan ($2.74 billion).

6　This hefty investment reflects the company's vision of VR as a key player in the future of technology at a time when the metaverse was the hottest concept out there. In 2021, Facebook rebranded as Meta, Disney officially established a metaverse division, and major Chinese companies all formed teams related to XR and the metaverse. But things didn't go according to plan.

7　ByteDance had hoped that an upgraded VR headset, a high-cost performance PICO 4, could help it establish a dominant market position, similar to what Meta's Quest headset was able to achieve. The company set an ambitious target of selling 1 million units for 2022, doubling on Pico's half million in total sales in 2021. However, despite leveraging its extensive short video platforms for marketing and enlisting numerous celebrities for promotions, Pico's sales fell short. The total sales for 2022 were estimated to be around 700,000 units, well below their target. Sales dropped further in 2023, when the total AR/VR headsets shipments in the entire Chinese market slumped to only 328,000 units. With Pico's 59% market share, that means Pico may have sold around 194,000 units during the first six months of 2023.

Consumers Find VR Headsets Aren't Must-Haves

8　To be sure, the design and performance of the PICO 4, priced at 2,500 yuan ($343), offers great value. But consumers are still not convinced that it's something essential to have in their lives. "After playing it for a few times, it was left on the shelf collecting dust," noted one Pico user.

9   The problems are manifold: high price entry barriers, subpar user experiences, the lack of quality content, and the absence of "killer apps." Users still complain of practical issues, such feelings of dizziness, lags in the software, heavy headsets, short battery life, compatibility with glasses, and controller sensitivity.

10   A lack of compelling content is also hindering the market. The most popular VR experiences today remain old games like Beat Saber. Emerging VR sports games struggle to remain viable, and users often find traditional gaming consoles like the Switch more appealing. The Pico platform has only about 530 applications, including games, videos, sports, and office scenarios — far fewer than the millions of apps available on smartphones or the tens of thousands of titles available on video game consoles.

11   At the end of the day, unlike personal computers and smartphones, which are intrinsically tied to everyone's work and daily lives, VR headsets have not achieved the same level of integration. For now, VR remains primarily an entertainment device. That means it probably can't justify billions of dollars of expenditure. It's perhaps wise for ByteDance to avoid throwing good money after bad.

VR's Time Will Come, Just Not Now in China

12   Despite facing challenges, the VR industry is still expected to play a crucial role as a next-generation mobile terminal or computing platform. According to an IDC report, the VR/AR headset market is projected to grow by 46.8% in 2024, driven primarily by two major products: Meta's Quest 3 and Apple's Vision Pro.

13   However, the outlook for the Chinese market appears more cautious, especially with Pico and Tencent scaling back their efforts. There are no expectations for any major product releases in China in 2024. Chinese VR/AR headset manufacturers are likely to focus on sustaining their operations, waiting for a transformative moment akin to the iPhone "revolution" in the future.

The Forbes 2023-11-17

Comprehension Check for Reading 2

Direction: Skim the article and answer the following questions.

1. In what way is ByteDance and Tencent scaling back in VR market?

2. What is the pivotal juncture of the VR industry?

3. Did ByteDance achieve its ambitious selling target of upgraded VR headset in 2022?

4. Why can't VR headsets be as popular as personal computers and smartphones among the consumers?

5. What products will contribute most to the headset market in 2024 according to the passage?

Unit Nine

Internet of Things

第九单元

物联网

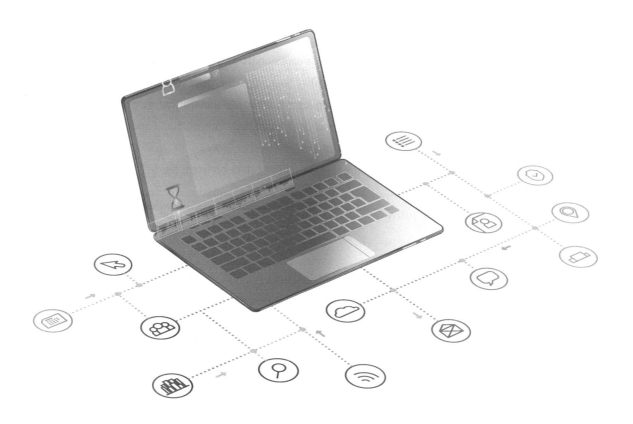

第一章 主题阅读

> Text

The Internet of Things: Applications for Business

What is IoT?

1 Defining IoT is hard for two reasons. Firstly, IoT has a breadth of applications; from monitoring supply chains to stopping trains and lighting homes. This wide range of uses means it is often hard to pinpoint exactly what unites the technologies grouped under IoT. Secondly, the technology is often referred to alongside Artificial Intelligence (AI) and Big Data as part of a triad at the centre of the Fourth Industrial Revolution (4IR). This is a relevant grouping, as IoT is a source of the Big Data needed to create AI algorithms(算法); however, this coupling has resulted in limited understanding of IoT itself.

2 For this research, we asked 20 industry professionals and academics how they define IoT. The recurring message focused on IoT's ability to bridge the digital and material worlds. Therefore our definition hinges on the following aspect of IoT's transformative potential: The Internet of Things is a network of physical objects or devices that communicate and interact with each other via an internet connection. The other central component of IoT is the internet connection. Connectivity is necessary for the transmission of data between the IoT object and the computing power that is collecting and analysing the information. Telecommunications (telecoms), as the sector that provides internet connections, is therefore at the heart of IoT transformation.

What can IoT do?

3 IoT can revolutionise the business and consumer landscape by bridging digital and material worlds. Any industry reliant on making, moving or selling objects that were previously not connected to the internet stands to benefit. The specific benefits IoT can bring to a business depend on how the technology is used. For example, sensors can be used to reduce waste by optimising lighting or heating based on occupancy levels, or reduce spoilage of products in transit by monitoring temperatures. IoT can also generate revenue and increase productivity, such as acoustic offshore oilfield sensors that analyse activity through pipelines to maximise output and help identify new resource pools. To help build an understanding of the potential for IoT in any given business, it is useful to consider the five key capabilities of IoT: connecting, collecting,

monitoring, monetising and optimising.

4 Connecting – IoT allows all manner of devices to become integrated and connected, moving them out of their respective silos and bridging the gap between the digital and the physical or "real" world. As the digital footprint of devices expands, the global system of connectivity becomes more robust and responsive to change.

5 Collecting – Sensors are a core component of IoT technology. They collect data from the object they are placed on, which can be used to inform other functions. The data collected also has value. Companies can aggregate, anonymise (匿名化) and sell data to interested third parties.

6 Monitoring – In its role as aggregator, IoT facilitates the ability to engage in remote monitoring, providing a rich, detailed snapshot of the world as it stands in real time. Existing compliance and monitoring of processes and assets can be automated and made more efficient through preventative (预防性的) and predictive (预设的) applications of IoT.

7 Monetising – IoT allows companies and sectors to become data-rich, both in terms of collecting and analysing data, sometimes from unexpected or hard to reach places. These new data streams bring with them significant opportunities for new revenue streams, either through aggregation or anonymisation or through adding new functionality that can be sold to the consumer.

8 Optimising – New levels of efficiency can be attained through IoT data collection, providing potential cost, energy or time savings in a variety of sectors, from healthcare to energy.

Why is IoT important?

9 This breadth of applications means IoT is set to have a major impact on the global economy in the next five years. Some of this transformative potential is already being seen. For example, the consumer product market is already posting gains, thanks to the growing popularity of health and entertainment wearables (可穿戴的健康和娱乐设备) and smart homes. IoT solutions are estimated to have risen from US$72bn in 2015 to US$236bn in 2020. Improvements

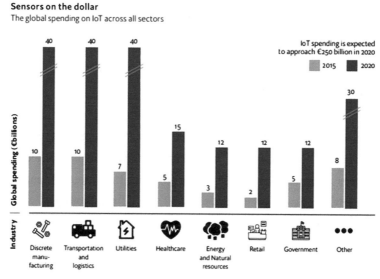

in natural language processing will see voice functionality further integrated into smart home tools; one forecast predicts that voice assistants will be integrated into eight billion domestic products by 2023. The smart home gets a lot smarter. Businesses that adopt IoT in their operations will become more competitive and new digital products will appeal to increasingly connected consumers. Those who do not adopt IoT may struggle as the technology increases the efficiency of competitors to meet the needs of their target markets. Awareness of the benefits IoT can bring is seen in the rise in spending on IoT. Approximately US$6trn will be spent on IoT solutions over the 2016—2021 and globally, IoT has fuelled more than US$80bn in merger and acquisition (M&A) investments, along with more than US$30bn in venture capital.

The Internet of Things: Applications for Business (The Economist Intelligence Unit Limited) 2020

Vocabulary

acoustic *adj*. 听觉的；原声的；
appeal *v*. 呼吁；上诉；对…有吸引力
application *n*. 申请；请求；申请书；运用；应用软件
bridge *v*. 弥合，消除；在……上架桥
capability *n*. 才能，能力；能量，性能
define *v*. 阐明；限定；给……下定义
facilitate *v*. 促进；使便利
gain *n*. 增加；好处；利润
group *v*. 使成群，集合；归类
hinge *n*. 铰链，合叶；关键，转折点；枢要，中枢

landscape *n*. 风景；形势
merger *n*. 合并；联合体；吸收
monetise *vt*. 定为货币，铸造成货币
monitor *v*. 监控；监听 *n*. 监测仪器；显示屏；班长；监督员
occupancy *n*. 占有，占领；居住
range *n*. 一系列；区间，范围
robust [rəu'bʌst] *adj*. 强健的；结实的；强劲的；坚定的
transmission *n*. 播送；传送

Cultural Notes

Fourth Industrial Revolution It represents a fundamental change in the way we live, work and relate to one another. It is a new chapter in human development, enabled by extraordinary technology advances commensurate with those of the first, second and third industrial revolutions. These advances are merging the physical, digital and biological worlds in ways that create both huge promise and potential peril. The speed, breadth and depth of this revolution is forcing us to rethink how countries develop, how organisations create value and even what it means to be human.

Read between Lines

Direction: Read the passage carefully and answer the questions below.

Question 1: Why is it difficult to define IoT? (Para.1)

_____.

Question 2: How does the professionals define the Internet of Things according to the passage? (Para.2)

_____.

Question 3: Which component is essential to the IoT technology? (Para.5)

_____.

Question 4: What capability of the IoT do you think is the most important one? Why do you think so? (Para.4~8)

_____.

Question 5: Why does the author insist that the IoT has the potential to influence the world economy? List examples to prove. (Para.9)

_____.

Question 6: What writing technique has been adopted to prove the benefits IoT can bring? (Para.9)

_____.

Words and Expressions

Ex.9.1 *Write the word according to the definition. The first letter is given.*

1. f_____ : make easier; be of use.
2. c_____ : the quality of being capable.

3. b_____: make a bridge across; connect or reduce the distance between.
4. r_____: sturdy and strong in form; strong enough to withstand challenges or adversity.
5. m_____: keep under surveillance.
6. g_____: the amount by which the revenue of a business exceeds its cost of operating.

Ex.9.2 *Fill in the blanks of the appropriate word from the box below. Change the form if necessary.*

group	robust	monitor	define	merger
potential	appeal	application	fuel	transmit

1. The law stipulates modes for company _____, bankruptcy, dissolution and liquidation (清算).
2. If they detect something wrong while the Wireless Security System is armed, they'll _____ a wireless alert signal to a base station that will then raise the alarm.
3. The books are _____ together by subjects in the library.
4. Various inducements aim to bring employees' _____ and creativity into full play.
5. Dress designing _____ to me.
6. This _____ domestic demand, pushing up pressure on prices, particularly of commodities.
7. She submitted a job _____ yesterday and the company asked her in for an interview today.
8. More women than men go to the doctor. Perhaps men are more _____ or worry less?
9. If you are a college student looking for work but worry you won't have enough time to devote to academic subjects, consider working as a study hall or library .
10. Suburbs are _____ as those areas that lie outside cities and towns in most dictionaries.

Ex.9.3 *Complete each sentence below with the appropriate form of word in bracket.*

1. He is a manager _____ (capability) of leadership.
2. There is no general agreement on a standard _____ (define) of intelligence.
3. Unfortunately, the current low prices for oil, gas, and coal may provide little incentive for research to find even cheaper substitutes for those _____ (fuel).
4. Heterosexual contact is responsible for the bulk of HIV _____ (transmit).
5. Penguins, like other seabirds and marine mammals, _____ (occupancy) higher levels in the food chain and they are what we call bio-indicators of their ecosystems.
6. The invention would have a wide range of _____ in industry. (application).
7. He was arrested for _____ (facilitate) the sale of alcohol to minors.
8. In the early 20th century, few things were more _____ (appeal) than the promise of scientific knowledge.

9. _____(merger) and acquisitions (M&As) reflect strategic transformation in traditional conception of competition.
10. The companies merged to _____(optimize) the allocation of resources.

Ex.9.4 *Translate the following Chinese sentences into English.*

1. 物联网能够联通数字、虚拟的世界和实体世界。(bridge)

_____.

2. 传感器可以通过检测温度而减少运输过程中对产品的破坏。(transmit)

_____.

3. 是否过度使用互联网不应该用上网时间来定义。(define)

_____.

4. 大学教育的目的之一就是推动学生自主学习。(facilitate)

_____.

5. 西方战略家(strategist)认为中国具备成为世界超级大国的潜力。(potential)

_____.

Theme Writing

Directions: You are required to write a passage of about 120 words based on your understanding of the short passage below. Your writing should respond to the following questions:

1) Do you think modern technology has caused the loss of privacy?
2) How should we protect our privacy in modern society?

  Privacy and security concerns have also been identified as relevant risks for the human-robot context (Mitzner et al., 2017). A study exploring opinions of televideo technology, including robot platforms such as the Beam, found that participants had, "concerns about invasion or breach of privacy, exposure of one's personal information/surroundings/life…" (Mitzner et al., 2017, p. 7). For example, it has been a major concern for Facebook users ever since the network's beginning in 2004.

第二章　拓展阅读

Reading 1

China Standard Plan 2035

1 The necessity to provide standards in the innovation technology sector, and more specifically in IoT, has been highlighted in the China Standards Plan 2035. This Plan will drive the Chinese government and leading technology companies to set global standards for emerging technologies, including IoT, 5G internet, and artificial intelligence, among others.

2 The China Standards Plan 2035 – to be read together with the Made in China 2025 strategy – demonstrates China's intention to become a leader in tech innovation, with the aim of achieving technology self-sufficiency and fostering the ability of Chinese companies designing and producing high-tech products as well as setting the relevant standards.

3 Setting technical standards holds critical value – it enables interoperability (协同工作的能力) between devices and their usability worldwide. Since international panels have long been dominated by Western countries – China is keen to change the status quo in future. Tech giants like Nokia Corp. and Qualcomm Inc. earn billions in revenue annually from the patents used by rivals making smartphone systems – and going forward, Beijing wants to ensure that standards are designed to align with the technologies developed by its own enterprises.

4 A recent piece published by The Wall Street Journal reported on China's "channeling of state funding and political influence" to take the driving seat in organizations that set up technical standards for cutting-edge technologies, especially those reliant on 5G networks. From the paper: "Chinese officials direct at least four global standards organizations; official documents estimate Beijing and regional governments supply annual stipends (薪俸) of up to 1 million yuan (roughly US$155,000) for firms leading global standards development, while Western funding is dwindling."

5 Recent information from the Ministry of Industry and Information Technology (MIIT) revealed that 90 domestic Chinese companies had filed a joint application to establish the National Integrated Circuit Standardization Technical Committee with its proposed secretariat at the China Electronics Standardization Institute. This is a relatively weak area for China as the

industry is dependent on access to US technology and is populated by small-scale enterprises. The 90 Chinese firms include Huawei, HiSilicon, Xiaomi, Datang Semiconductor, Unichip Microelectronics, Zhanrui Communication, ZTE Microelectronics, SMIC, Datang Mobile, China Mobile, China Unicom, ZTE, Tencent. Through the committee, these firms intend to strengthen cooperation between weak industries.

Incentivizing Foreign Investments in Tech Innovation

6 Last year, China's new infrastructure plan was developed into a strategy to meet the twin goals of stimulating job creation and preparing for new changes in the global economy, particularly in the realm of technology and sustainable development. Foreign investment opportunities were carved out in state policies and incentive schemes for business stakeholders to participate in this next phase of China's development with key segments getting liberalized like industrial internet, new energy vehicles ("NEV"), and artificial intelligence.

Find Business Support

7 In particular, China rolled out a wide range of favorable policies for the integrated circuit (IC) and software industries – tax breaks, favorable financing, IP protection, and support for R&D, import and export, and talent development etc. to attract foreign know-how. This has also been followed by amendments to its legal regime to strengthen intellectual property protection safeguards.

8 Among other industries where foreign companies are encouraged to invest in China, it is worth mentioning the opening of the automotive industry, including the manufacturing of NEVs. Joint venture ("JV") restrictions for manufacturing passenger automotive vehicle – in terms of equity ratio and maximum number of JVs that can be established by a foreign investor in China – shall be nationally removed in year 2022.

9 Here it should be noted that foreign investment into information communication technology (ICT) – where IoT is applied – is allowed but foreign companies often face difficulties in securing licenses and meeting other establishment requirements; many find that regulations relating to data management and cyber infrastructure, for instance, favor domestic firms. That is why they often choose the JV route to enter the Chinese market; going forward, these hidden barriers to direct and free market access may need to be addressed.

China Briefing 2021-2-18

Comprehension Check for Reading 1

Direction: Skim the article and answer the following questions.

1. Which of the following is not relevant to the purpose of the China Standards Plan 2035?

A) To become a leader in tech innovation.

B) To foster abilities of setting standards for Chinese government.

C) To achieve technology self-sufficiency.

D) To foster the ability of Chinese companies designing and producing high-tech products.

2. Why does the author insist setting technical standards holds critical value for China?

A) Because it has long been dominated by Western countries.

B) Because corporations like Nokia Corp. and Qualcomm Inc. earn billions in revenue annually from mobile phones.

C) Because China intends to develop the technology dominated by the West, which requires related standards.

D) Because they involve interoperability which is key to the success of China.

3. Which effort is not true for Chinese officials and companies to make for the standardization development?

A) Chinese governments have financed 1 million yuan for its development.

B) Chinese home companies have agreed on a committee for the establishment of the standardization.

C) Chinese small-scale enterprises have decided to keep depending on US technology.

D) The cooperation in weak industries are expected to facilitate the 90 domestic companies' development.

4. Chinese government has taken all the following actions to support its software industries, except for which one?

A) To legally strengthen intellectual property protection safeguards.

B) To promote tax-free policymaking in import and export.

C) To sponsort researches and development.

D) To work out favorable policies for talent development in relevant fields.

5. Why do foreign investments tend to choose JV to enter Chinese market?

A) Because regulations relating to national security, like data management and cyber infrastructure, are intended to favor domestic firms.

B) Because foreign companies often face difficulties unable to be addressed later.

C) Because home companies can always enjoy priorities.

D) Because hidden barriers to direct and free market access may need to be addressed.

Reading 2

China Releases 10-year Vision, Action Plan for BRI, Focusing on Green, Digital Development and Supply Chain

1 China on Friday published a document entitled "Vision and Actions for High-Quality Belt and Road Cooperation: Brighter Prospects for the Next Decade," which specified the key areas and directions for Belt and Road cooperation in the next 10 years.

2 The China-proposed Belt and Road Initiative (BRI) has injected new growth momentum into the world economy and created vast room for global development. The initiative is believed to bring benefits to its partner countries through the newly released vision and action plan for BRI's development in the next decade, with some new focuses on green and digital development, as well as the stability and sustainability of supply chains.

3 From a China-proposed initiative to an international practice, the BRI, which celebrates its 10th anniversary in 2023, has become a well-received international public good and a platform for international cooperation.

4 Analysts noted that the new action plan comes in time for future development, as the next "golden decade" for BRI starts and the plan involves new industries and global concerns. New cooperation sectors such as innovation, the digital economy and green development should be explored to inject vitality and momentum into the BRI, they said.

New focuses

5 In addition to what the BRI has been devoted to in the past decade — policy coordination, infrastructure connectivity, unimpeded trade, financial integration and people-to-people ties — the Friday document added cooperation in new fields: green development, new forms and models of digital cooperation, technology innovation, international cooperation in health, according to the document.

6 The new action plan will also focus on promoting the organic integration of trade and the latest technologies, including internet, Internet of Things, big data, artificial intelligence (AI) and blockchain.

7 There is vast opportunity for digital cooperation between China and BRI partner countries, as China has extensive experience in digital infrastructure construction and digital development. At the same time, China has a huge number of digitalized industries, Wang Peng, an associate researcher with the Beijing Academy of Social Sciences, told the Global Times on Friday.

8 "In terms of AI, China and BRI partners can strengthen cooperation and exchanges on AI governance and rules, and research and development of AI technology," said Wang.

9 During a Friday press conference that outlined the new BRI action plan, Chinese officials also said that making industrial and supply chains more resilient and expanding the scope of free trade agreements will be a focus of BRI cooperation in the future.

10 "What we are advocating is mutual benefit and win-win. Therefore, China plays a very important role in the development of global trade, as well as the stability of industries and supply chains. In particular, global trade protectionism has been prevalent in recent years, but China has always been committed to opening-up," said Hu Qimu, deputy secretary-general of the digital-real economies integration Forum.

11 In the past 10 years, the total import and export volume of China and BRI partner countries has reached $19.1 trillion, and two-way investment has exceeded $380 billion, according to statistics from the National Development and Reform Commission (NDRC), the country's top economic planner.

Well prepared

12 The BRI has gradually progressed from an idea into actions and from vision to reality by yielding fruitful results to countries and people around the world. Now China aims to promote high-quality, sustainable and people-centered development of the BRI through joint efforts to bring brighter prospects for the next decade, Maya Majueran, director of Belt and Road Initiative Sri Lanka (BRISL), a Sri Lanka-based organization that specializes in BRI cooperation, told the *Global Times* on Friday.

13 "No doubt that BRI will bring more and more benefits to partner countries and wealth and strength to the Global South and change the geopolitical balance between East and West," said Majueran.

14 China is well prepared to promote the high-quality development of BRI in the future.The third Belt and Road Forum for International Cooperation held in October produced 458 deliverables, and Chinese financial institutions established a financing window of 780 billion ($109 billion) for BRI projects. All this will contribute to high-quality BRI cooperation, and provide strong momentum for connectivity, development and prosperity around the world, an official said on Friday at the press conference.

Open to all

15 At present, China has signed more than 200 documents with 152 countries and 32 international organizations on cooperation under the BRI, covering 83 percent of the countries with which China has established diplomatic relations, according to official statistics.

16 The Gulf Cooperation Council has been one of the biggest beneficiaries of the BRI, Hazem Ben-Gacem, CEO of InvestCorp, the largest non-sovereign wealth fund private equity platform in the Middle East, told the *Global Times*.

17 "The mutually beneficial and win-win concept of the BRI is greatly appreciated and respected by investors from the Middle East," Ben-Gacem said.

18 Majueran noted that the West is jealous of BRI and accused China of engaging in "debt trap" diplomacy to discredit the initiative.

19 "A number of latest studies including Western studies show that there is no evidence of a so-called debt trap. The West proposed multiple initiatives to counter the BRI such as Partnership for Global Infrastructure and Investment, but they were all talk with no action," Majueran noted.

20 China's move is in stark contrast to the US-led Western world's "small yard, high fence" approach of walling itself off from global cooperation, Bao Jianyun, director of the Center for International Political Economy Studies at Renmin University of China, told the Global Times.

Comprehension Check for Reading 2

Direction: Skim the article and answer the following questions.

1. Which of the following is true about BRI?

   A) The Belt and Road Initiative was proposed by 152 countries.

   B) It has been established for less than 10 years till 2023.

   C) It focuses on green and digital development, the stability and sustainability of supply chains.

   D) It has witnessed the "golden decade" in the past 10 years.

2. Which one does not belong to the positive roles China has played in fulfilling the win-win situation for countries of BRI?

   A) China has devoted itself to the opening-up.

   B) China attaches importance to the supply chains.

   C) China has made efforts to play a part for the stability of industries.

   D) China once followed the trend of trade protectionism.

3. According to Maya Majueran, _____ is not mentioned under the influence of BRI.

   A) participating countries of BRI.

   B) global South.

   C) the balance between Easter and Western countries.

   D) US-led Western world.

4. Which is not the result with the efforts of the third Belt and Road Forum and Chinese financial institutions?

   A) High-quality cooperation for BRI is expected to achieve.

B) They have delivered 19.1 trillion products in the past 10 years.

C) They will contribute to the realization of the world's connectivity, development and prosperity.

D) A financing window worth more than 100 billion for BRI was established.

5. Which one is right about the West's attitude towards BRI?

A) They are talking big.

B) They are malicious to China.

C) They are appreciative of its concept.

D) They have shown jealousy to it.

Unit Ten
Data Security

第十单元
数据安全

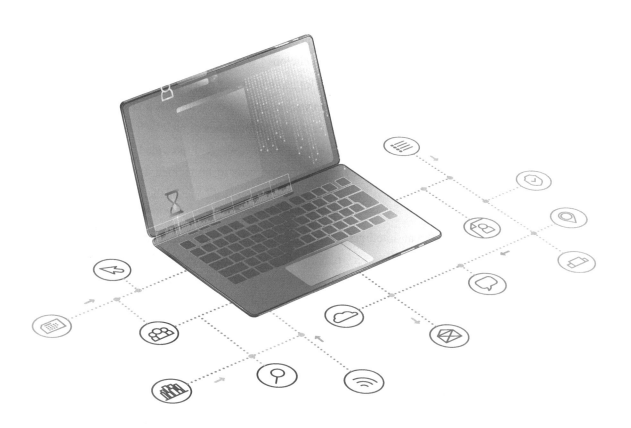

第一章　主题阅读

> **Text**

Cybercriminals Are Now Targeting Top Executives and Could Be Using Sensitive Information to Extort Them

1 The workcation is on the rise. The blended work-travel experience is seen by proponents as a great way to increase motivation, productivity, and creativity. In a recent YouGov poll, 53% of Americans who can work remotely said they were interested in taking a workcation in the next 12 months - and their companies are helping make this happen.

2 New ways of working have extended to business leaders. Around 80% of executive jobs are currently available remotely, compared to 25% pre-pandemic, according to executive search firm Cowen Partners. However, senior executives need to be the most careful, whilst having the chance to reflect, recharge, and reset. If the upper echelons (等级，阶层) of management choose to work away from home and the office, there is a lot more at stake for a business.

3 According to this year's Cybersecurity Breaches Survey from the UK's Department for Science, Innovation & Technology, 32% of businesses recall breaches or attacks from the last 12 months. The report further found that only 18% of businesses had provided cybersecurity training to its workforce. Addressing the cyber skills gap needs to be tackled across all levels, including the C-suite, especially when they are the prime targets of one of the fastest-growing forms of cyber threats.

4 Our Talos global threat intelligent team has identified a 25% increase in data theft extortion incidents in the latest quarter, making it the most observed threat, overtaking ransomware (勒索软件). Data theft extortion is a cybersecurity incident where personal or confidential data is stolen. Once access is gained and data is analyzed, the bad actor then threatens to release sensitive information unless the victim meets their demands and pays out, which typically involves the transfer of cryptocurrency (加密货币). By choosing victims at the top of the tree, the value of the data is no doubt perceived to be higher and hackers see the potential to obtain payments well into the millions.

5 For the business leader, having access to reliable Wi-Fi is far from the only priority whilst traveling abroad. The importance of maintaining strong cyber hygiene practices when they are connecting to the corporate network from locations outside their home turf (势力范围)

becomes even more critical. Because unlike them, cybercriminals will not be taking a holiday. Fortunately, security measures to protect senior executives, and ultimately their business, need not be difficult.

6 Believe it or not, the most common passwords of 2022 still included the likes of "123456," "QWERTY," and "password." Most cyberattacks and data breaches remain the result of weak passwords. Ensuring an executive's identity remains their own requires a more complex approach to passwords that are longer, stronger, and harder for someone else to guess. Unique passwords for each account should use a variety of cases and symbols and be changed annually.

Defending Devices Abroad

7 Business leaders abroad need to be extra wary of equipment theft as one of the leading causes of data theft is device loss. It is also important to make life more difficult for anyone trying to snatch a device, particularly mobile phones. A cybercriminal who gets hold of an online password can sign into an account from anywhere, but if they obtain the PIN, they can access the device too. As research has shown, the choice of PIN or password determines how vulnerable devices are to a successful hack or attack, and obvious identifiers such as a birthday or credit card PIN should be avoided. Apps should only be installed from trusted sources and software should be kept up to date.

Authentication at Home and Away

8 Two-factor authentication (or 2FA) adds an extra layer of security that is easy for remote bosses to use and makes it significantly more difficult for anyone who should not be accessing their data to do so. 2FA technologies mean the executive has the chance to catch malicious abuse of their credentials and administrators have the metadata from the user's acknowledgment, so they can see any unusual time or location of access.

Endpoint Protection Everywhere

9 Having a virtual private network (VPN) is another important security solution for business leaders. An encrypted connection over the Internet from a device to a network helps ensure that sensitive data is safely transmitted, preventing unauthorized people from eavesdropping on data traffic. Users can then use the virtual network to access the corporate network without risk.

10 As business leaders may extend their holidays to make the most of time away from their office desks, it is even more critical that they follow good cybersecurity practices to prevent the risk of a serious attack.

11 With the freedom and power of hybrid working also comes the responsibility of senior bosses to work in a cyber-secure manner, wherever they are. A more relaxed environment shouldn't mean a more relaxed approach to security, especially when it's at the expense of their business.

The Fortune: Cybercriminals are now targeting top executives – and could be using sensitive information to extort them   2023-9-13

Vocabulary

address *n.* 地址，网址；演说 *v.* 写 (收信人) 姓名地址；解决，处理；演讲；称呼

blend *v.* (使) 混合；融合，结合；协调 *n.* 融合；混合 (物)

breach *n.* 违反，破坏；破裂，缺口 *v.* 违反，破坏；突破，攻破

confidential *adj.* 机密的，保密的；悄悄的

eavesdrop *vi.* 偷听 (别人的谈话)

encrypt *v.* 加密，将…译成密码

executive *n.* (公司或机构的) 经理，领导层；(政府的) 行政部门 *adj.* 行政的；管理的；高级的

extortion *n.* 敲诈，勒索；被勒索的财物；敲诈者

hygiene *n.* 卫生，卫生学；保健法

identify *vt.* 识别，认出；确定；使参与；把……看成一样

malicious *adj.* 恶意的，有敌意的；蓄意的；预谋的；存心不良的

proponent *n.* 支持者，拥护者，提倡者

recharge *v.* 再充电；再装填 (弹药等)；再控告；休整，养精蓄锐

snatch *n.* 抢；一阵子，很小的数量 *vi.* 作出握住或抢夺的动作；很快接受 *vt.* 抢夺，夺得；及时救助

wary *adj.* 小心的，谨慎的

Cultural Notes

YouGov poll It is a global public opinion and data company founded in 2000 as a pioneer of polling and the most quoted pollster in the world.

Cowen Partners An top executive search firm hat specializes in identifying and placing exceptional leaders in key positions for companies of all sizes and industries.

Department for Science, Innovation & Technology (DSIT) It aims to position the UK at the forefront of global scientific and technological advancement; to drive innovations that change lives and sustain economic growth; to deliver talent programmes, physical and digital infrastructure and regulation to support British economy, security and public services, R&D funding.

2 FA Technologies 2 FA is an identity and access management security method that requires two forms of identification to access resources and data. 2FA gives businesses the ability to monitor and help safeguard their most vulnerable information and networks. Businesses use 2FA to help protect their employees' personal and business assets.

Read between Lines

Direction: Read the passage carefully and answer the questions below.

Question 1: What is the possible meaning of "workcation"? (Para.1)

_____.

Question 2: Why do senior executives need to be the most careful according to Para.3? (Para.3)

_____.

Question 3: Which is more serious, data theft extortion or ransomware? (Para.4)

_____.

Question 4: How do criminals usually manipulate in a data theft extortion?(Para.4)

_____.

Question 5: Which methods are listed as important security solutions to data theft extortion by the author? (Para.5～9)

_____.

Question 6: What benefits can senior bosses gain from workcation? As return, what must be paid attention to ? (Para.10～11)

_____.

Words and Expressions

Ex.10.1 *Write the word according to the definition. The first letter is given.*

1. b_____: act in disregard of laws and rules.
2. w_____: marked by keen caution and watchful prudence.
3. b_____: combine into one.

4. m_____: wishing or appearing to wish evil to others.
5. a_____: direct one's efforts towards something.
6. v_____: being such in essence or effect though not in actual fact.

Ex.10.2 *Fill in the blanks of the appropriate word from the box below. Change the form if necessary.*

| Identify | address | confidential | blend | recharge |
| Wary | snatch | malicious | extortion | executive |

1. One sneaky thing some malware _____ threats does is to modify a user's server information.
2. Gardens of Suzhou _____ the humane arts with nature, creating an atmosphere of living in harmony.
3. The kidnappers _____ a £75,000 ransom form her parents for her release.
4. In our world today, _____ information is the lifeblood of any business.
5. The American came from behind to _____ victory by a mere eight seconds.
6. We need very broad participation to fully _____ the global tragedy that results when countries fail to take into account the negative impact of their carbon emissions on the rest of the world.
7. Scientists have _____ a link between diet and cancer.
8. Hot springs are perfect for people who want to relax and _____ themselves after a long week.
9. People have found it necessary to teach their children to be _____ of strangers who offer them gifts.
10. The President of the United States is the _____ head of the government.

Ex.10.3 *Complete each sentence below with the appropriate form of word in bracket.*

1. The wolves _____ (wary) approached the camp fire looking for food.
2. Most of digital satellite TV systems are set_____ (remote) .
3. Now that the problem has been _____(identify), appropriate action can be taken.
4. Any details provided will be treated with _____(confidential) and will not be made public.
5. Their actions of hegemonism threatened a serious _____(breach) in relations between the two countries.
6. Officials emphasized the importance of individual care and good _____ (hygiene) practices during the pandemic.
7. A large number of creative new retailers, like Bonobos, are _____(blend) online and

offline experiences in creative ways.
8. Facing an 8.3-billion-dollar budget deficit in 2016, the Postal Service close one of several _____ (proponent) that has put forth recently to cut costs.
9. If _____ (executive) fail to exploit the opportunities of networking, they risk being left behind.
10. Though _____ (malicious) may darken truth, it cannot put it out.

Ex.10.4 *Translate the following Chinese sentences into English.*

1. 网络技能的空白需要在各个层面加以解决。(address)

_____.

2. 抄袭别人的研究成果严重违反道德。(breach)

_____.

3. 掌握在线密码的网络罪犯可以从任何地方登录帐户，但如果他们获得了 PIN，也可以访问该设备。(password, sign in)

_____.

4. 在国家安全问题上，我们需要提高警惕。(wary)

_____.

5. 在计算机技术领域中，保护用户数据的机密性是至关重要的。(confidentiality)

_____.

Theme Writing

Directions: You are required to write a passage of about 120 words. Your writing should respond to the following questions:

Many people do not feel safe in the information era either at home or they are out. What are the causes, and what can be done to make people feel safer?

第二章 拓展阅读

Reading 1

How Didi Crashed Into China's New Data Security Laws

1   On June 10, 2021, China's Standing Committee of the National People's Congress passed the Data Security Law (DSL). The law has been under review since June 2020 and aims to further strengthen regulation on data collection, storage, and distribution across China's rapidly-growing digital economy. It stipulates a top-down coordination of data security practices and raises the stakes for compliance with strict fines and punishments for violators. Amidst an unprecedented regulatory crackdown that has defined the year for a wide range of Chinese industries, this law represents a broader transition for policymakers from reigning in anti-competitive practices to addressing data handling and security.

2   The DSL mainly focuses on two main tenets regarding the usage of data in China. Firstly, it categorizes sensitive data about national security and stipulates different data storage and exportation requirements by tier. Secondly, it advocates for anti-monopolistic collection and usage of data. In the case of Chinese ride-hailing giant Didi Chuxing Technology Co. ("Didi"), the firm's data practices raised both flags, which led to a serious investigation that culminated in sweeping penalties. The Cyberspace Administration of China ("CAC") and the State Administration for Market Regulation ("SAMR") argued that Didi was anti-competitively using data for its own gain, and regulators noted that the firm's offshore IPO in the United States could lead to the exportation of nationally sensitive data. As a result, regulators delisted the Didi app from app stores, barred new user registrations, and are planning a multi-billion dollar fine for the company. The case has drawn international attention, and, for those doing business in China, marks the beginning of a new compliance environment in China.

What Are China's Data Security Laws?

3   At the moment, there are three laws related to data and information protection in China: The Cybersecurity Law of the People's Republic of China ("Cybersecurity Law"), the Data Security Law ("DSL"), and the Personal Information Protection Law of the PRC ("PIPL"). The Cybersecurity Law was instituted in mid-2017, the DSL took effect as of September 2021, and the PIPL is set to launch in November 2021. Together, these three laws aim to build a comprehensive legal framework for data regulation.

What Does the Data Security Law Cover?

4　The Data Security Law is the only law of the three to focus exclusively on data and the companies and individuals that process it. Building on the framework of its predecessor, the Cybersecurity Law, the DSL continues on to construct a system that regulates nationally sensitive data. Together, the Cyberspace Administration of China and the National Security Commission now oversee the nationally standardized categories for data classification and have built systems to collect and assess the data-associated risks of organizations and individuals.

5　While the objective seems clear cut, the business implications are anything but. In fact, the law itself lacks a clear definition of what "nationally sensitive data" is. For businesses, the law threatens heavy fines for failure to comply with "national core data" rules, which include exporting data to foreign authorities, failing to comply with data requests, or failing to fulfill data security obligations. In an all-too-familiar move, the DSL also makes specific mention of anti-competitive or illegal uses or collections of data, echoing sentiments from China's Anti-Monopoly Law.

China's Regulatory Crackdown and Data Security

The Rise of Anti-Monopoly Regulation

6　Currently, the Chinese government is in the midst of an unprecedented crackdown on some of the biggest internet companies in the world. The State Administration for Market Regulation (SAMR) has recently imposed a record fine on e-commerce giant Alibaba, initiated an antitrust probe into food delivery company Meituan, and levied fines against a multitude of other industry titans. Following the release of proposed changes to China's Anti-Monopoly Law in early 2021, SAMR ordered these companies to conduct self-inspections for any practices that violate market regulation and warned of severe consequences for companies that did not adjust accordingly.

7　While these rules are severe, they mainly tackled the monopolistic practices that plague the Chinese economy. For example, regulators fined Alibaba 18 billion yuan (US$2.78 billion) in April 2021 after alleging that Alibaba had abused its market position by preventing merchants from using other e-commerce platforms. Similarly, SAMR charged Tencent with employing various anti-competitive practices across some of its business arms, notably including its exclusive control over music streaming services in China. Didi Chuxing, China's ride-hailing giant, also fell victim to regulators as SAMR investigated whether the company attempted to squeeze out smaller ride-share rivals through anti-competitive pricing and marketing tactics.

What Is Driving China's Clampdown on Didi and Data Security?

8　Across the board, these fines and restrictions have targeted anti-competitive practices. However, following the release of China's Data Security Law, regulators began a new crackdown which focused more on corporate misuse of consumer data. This placed Didi at the center of regulatory crosshairs.

9 Nearly a month after the Data Security Law was passed in September 2021, officials from seven Chinese government departments visited Didi's offices in Guangzhou to conduct a cybersecurity review. The CAC, SAMR, and five other departments visited Didi for a network security review after the CAC alleged that the ride-hailing company had illegally collected users' data. Amid the review, Didi was required to stop new user registration and remove its app from Chinese app stores. This sent company shares failing by over 7%. Once hailed as the most popular IPO listing of 2021, the sudden crackdown dashed Didi's hopes for a smooth cash influx from US markets.

10 Another key risk to regulators was Didi's overseas IPO. Given the expansive nature of Didi's operations, the firm maintains a wide variety of Chinese user data, including user phone numbers that are linked to real names and identification. In the event of an IPO, Didi would likely have to comply with SEC policy that requires this data to be handed over for compliance audits. For Chinese regulators worried about national security and data leaks, this posed a significant threat should foreign governments have access to this information.

What Are Didi's Punishments?

11 In response to these violations, regulators weighed a wide range of punishments, including fines, operational suspensions, and the introduction of state-owned investors. Additionally, there was also consideration of a possible forced delisting of Didi from US markets.

12 Months in, regulators are still deciding Didi's punishment and the final ruling is still uncertain. However, it is expected that Didi will face a fine similar to the record smashing US$2.78 billion penalty imposed on Alibaba. In addition, the CAC and SAMR are also expected to require that Didi hand over control of its data to another domestic auditor, such as a data security company owned by a state-owned enterprise. This body would then be able to access Didi's servers across the entire country and track the company's data collection, usage, and transfers operations.

China Guys 2021-10-28

Comprehension Check for Reading 1

Direction: Skim the article and answer the following questions.

1. Which one does not belong to the significance of the newly-published Data Security Law?
   A) Coordination of data security practices is regulated.
   B) A transition from anti-competitive practices to data handling and security handling for policymakers will be ensured by the Law.

C) China's rapidly-growing digital economy will benefit from it.

D) Violators will be more severely punished based on the law.

2. What does the author mean by saying Didi Chuxing raised both flags?

A) It was delisted in the app store twice by Chinese regulators.

B) It was against two of the tenets in data practices by using data for its own profits and leaking sensitive data related to national security.

C) It was warned and punished twice for its anti-monopolistic collection and usage of data.

D) It resulted in a serious investigation and sweeping penalties.

3. Which one is not true about the three laws related to data and information protection in China?

A) Chinese governments have financed 1 million yuan for its development.

B) Chinese home companies have agreed on a committee for the establishment of the standardization.

C) Chinese small-scale enterprises have decided to keep depending on US technology.

D) The cooperation in weak industries are expected to facilitate the 90 domestic companies' development.

4. What is the author's attitude towards the law?

A) Positive.   B) Unclear.   C) Strict.   D) Negative.

5. Which is the chief reason for Chinese government to crackdown Didi according to CAC?

A) Because it has illegally collected and misused consumers' data.

B) Because it intended a cash influx from US markets.

C) Because it complied with SEC policy, regardless of national security.

D) Because its Chinese users' data include such privacy as real names and identification.

Reading 2

Data Security and GDPR

1 Data masking, data subsetting, and data redaction are techniques for reducing exposure of sensitive data contained within applications. These technologies play a key role in addressing anonymization and pseudonymization requirements associated with regulations such as EU GDPR. The European Union GDPR was built on established and widely accepted privacy principles, such as purpose limitation, lawfulness, transparency, integrity, and

confidentiality. It strengthens existing privacy and security requirements, including requirements for notice and consent, technical and operational security measures, and cross-border data flow mechanisms. In order to adapt to the new digital, global, and data-driven economy, the GDPR also formalizes new privacy principles, such as accountability and data minimization.

2 Under the General Data Protection Regulation (GDPR), data breaches can lead to fines of up to four percent of a company's global annual turnover or €20 million, whichever is greater. Companies collecting and handling data in the EU will need to consider and manage their data handling practices including the following requirements:

Data Security. Companies must implement an appropriate level of security, encompassing both technical and organizational security controls, to prevent data loss, information leaks, or other unauthorized data processing operations. The GDPR encourages companies to incorporate encryption, incident management, and network and system integrity, availability, and resilience requirements into their security program.

Extended rights of individuals. Individuals have a greater control — and ultimately greater ownership of – their own data. They also have an extended set of data protection rights, including the right to data portability and the right to be forgotten.

Data breach notification. Companies have to inform their regulators and/or the impacted individuals without undue delay after becoming aware that their data has been subject to a data breach.

Security audits. Companies will be expected to document and maintain records of their security practices, to audit the effectiveness of their security program, and to take corrective measures where appropriate.

What are the Challenges of Database Security?

3 Databases are valuable repositories of sensitive information, which makes them the primary target of data thieves. Typically, data hackers can be divided into two groups: outsiders and insiders. Outsiders include anyone from lone hackers and cybercriminals seeking business disruption or financial gain, or criminal groups and nation state-sponsored organizations seeking to perpetrate fraud to create disruption at a national or global scale. Insiders may comprise current or former employees, curiosity seekers, and customers or partners who take advantage of their position of trust to steal data, or who make a mistake resulting in an unintended security event. Both outsiders and insiders create risk for the security of personal data, financial data, trade secrets, and regulated data.

4 Cybercriminals have a variety of approaches they employ when attempting to steal data from databases:

Compromising or stealing the credentials of a privileged administrator or application. This is usually through email-based phishing, other forms of social engineering, or by using malware to uncover the credentials and ultimately the data.

Exploiting weaknesses in applications with techniques such as SQL injection or bypassing application layer security by embedding SQL code into a seemingly innocuous end-user provided input.

Escalating run-time privileges by exploiting vulnerable applications.

Accessing database files that are unencrypted on the disk.

Exploiting unpatched systems or misconfigured databases to bypass access controls.

Stealing archive tapes and media containing database backups.

Stealing data from nonproduction environments, such as DevTest, where data may not be as well protected as in production environments.

Viewing sensitive data through applications that inadvertently expose sensitive data that exceeds what that application or user should be able to access.

Human error, accidents, password sharing, configuration mistakes, and other irresponsible user behavior, which continue to be the cause of nearly 90% of security breaches.

Database Security Best Practices

5 A well-structured database security strategy should include controls to mitigate a variety of threat vectors. The best approach is a built-in framework of security controls that can be deployed easily to apply appropriate levels of security. Here are some of the more commonly used controls for securing databases:

6 **Assessment controls** help to assess the security posture of a database and should also offer the ability to identify configuration changes. Organizations can set a baseline and then identify drift. Assessment controls also help organizations to identify sensitive data in the system, including type of data and where it resides. Assessment controls seek to answer the following questions:

Is the database system configured properly?

Are patches up-to-date and applied regularly?

How are user privileges managed?

What sensitive data is in the database system? How much? Where does it reside?

7 **Detective controls** monitor user and application access to data, identify anomalous behaviors, detect and block threats, and audit database activity to deliver compliance reporting.

8 **Preventive controls** block unauthorized access to data by encrypting, redacting, masking, and subsetting data, based on the intended use case. The end goal of preventive controls is to halt unauthorized access to data.

9 **Data-specific controls** enforce application-level access policies within the database, providing a consistent authorization model across multiple applications, reporting tools, and database clients.

10 **User-specific controls** enforce proper user authentication and authorization policies, ensuring that only authenticated and authorized users have access to data.

Data Security Solutions

11 Reduce the risk of a data breach and simplify compliance with data security best

practices, including encryption, key management, data masking, privileged user access controls, activity monitoring, and auditing.

12 **Data protection:** Reduce the risk of a data breach and noncompliance with solutions to satisfy a wide range of use cases such as encryption, key management, redaction, and masking. Learn about Data Safe.

13 **Data access control**: A fundamental step in securing a database system is validating the identity of the user who is accessing the database (authentication) and controlling what operations they can perform (authorization). Strong authentication and authorization controls help protect data from attackers. Additionally, enforcing separation of duties helps to prevent privileged users from abusing their system privileges to access sensitive data, and also helps to prevent accidental or malicious changes to the database.

14 **Auditing and monitoring:** All database activity should be recorded for auditing purposes—this includes activity happening over the network, as well as activity triggered within the database (typically through direct login) which bypasses any network monitoring. Auditing should work even if the network is encrypted. Databases must provide robust and comprehensive auditing that includes information about the data, the client from where the request is being made, the details of the operation and the SQL statement itself.

15 **Securing databases in the cloud**: Cloud database deployments can reduce costs, free up staff for more important work, and support a more agile and responsive IT organization. But those benefits can come with additional risk, including an extended network perimeter, expanded threat surface with an unknowable administrative group, and shared infrastructure. However, by employing the proper database security best practices, the cloud can provide better security than most organizations have on-premises, all while reducing costs and improving agility.

Comprehension Check for Reading 2

Direction: Skim the article and answer the following questions.

1. About the improvements GDPR has made, which one is incorrect?
    A) They target for a new digital, global, and data-driven economy.
    B) Its purpose limitation, lawfulness and confidentiality have been changed.
    C) New privacy principles have been formalized.
    D) Existing privacy and security requirements have been strengthened.

2. What is the possible meaning of underlined "business disruption"?
    A) Interrupting a business negotiation.   C) Failure in businesses.
    B) Disturbing commercial activities.   D) Solving business disputes.

3. Which method can't be used to steal credentials for cybercriminals according to the author?

   A) Malicious softwares are made use of to reveal data.

   B) E-mails are commonly used to gain sensitive information by criminals disguised as someone else.

   C) Socializing parties can be important occasions to steal valuable information.

   D) Senior managers are often targets chosen by criminals.

4. Why does the author insist that built-in framework of security controls is the best approach?

   A) Because it includes various threat vectors.

   B) Because it is securer than any other approach.

   C) Because it is more commonly used to secure databases.

   D) Because it can be distributed scientifically and adopted for different security levels.

5. Which is not true as the primary to secure a database system?

   A) To provide robust and comprehensive auditing.

   B) To control what criminals can perform.

   C) To prevent hackers' changes to the database.

   D) To make the user valid legally.